Sprayed Concrete Lined Tunnels

Second Edition

Sprayed Concrete Lined Tunnels

Second Edition

Alun Thomas

CRC Press
Taylor & Francis Group
Boca Raton London New York

CRC Press is an imprint of the
Taylor & Francis Group, an **informa** business

CRC Press
Taylor & Francis Group
6000 Broken Sound Parkway NW, Suite 300
Boca Raton, FL 33487-2742

First issued in paperback 2021

ISBN 13: 978-1-03-204303-6 (pbk)
ISBN 13: 978-0-367-20975-9 (hbk)
ISBN 13: 978-0-429-26456-6 (ebk)

DOI: 10.1201/9780429264566

Library of Congress Cataloging-in-Publication Data

Names: Thomas, Alun, author.
Title: Sprayed concrete lined tunnels / Alun Thomas.
Description: 2nd edition. | Boca Raton : Taylor & Francis, a CRC title, part of the Taylor & Francis imprint, a member of the Taylor & Francis Group, the academic division of T&F Informa, plc, [2019] | Includes bibliographical references and index.
Identifiers: LCCN 2019010534| ISBN 9780367209759 (hardback) | ISBN 9780429264566 (ebook)
Subjects: LCSH: Shotcrete-lined tunnels.
Classification: LCC TA815 .T46 2019 | DDC 624.1/93--dc23
LC record available at https://lccn.loc.gov/2019010534

Visit the Taylor & Francis Web site at
http://www.taylorandfrancis.com

and the CRC Press Web site at
http://www.crcpress.com

To my parents for all their love and support.

To adapt Oscar Wilde's observation on socialism, the trouble with tunnelling is that it takes up too many evenings.

Contents

List of Figures

List of Tables

Preface

This book seeks to provide an introduction to the subject for people who have little experience of sprayed concrete lined (SCL) tunnels as well as serving as a reference guide for experienced tunnellers. Tunnels, for any civil engineering purpose, in both hard rock, blocky rock and soft ground are covered. In this context, blocky rock is defined as rock where the movement of blocks dominates behaviour. Soft ground is defined as soil or weak rocks where the ground behaves as continuous mass, rather than discrete blocks. The construction method is heavily influenced by the prevailing geology and in turn the method influences the design approach. Hence construction methods are discussed ahead of the sections on design in this book. That said, SCL tunnels arguably involve a much more interactive consideration of the construction method and the design than other tunnels. This is an iterative process.

Opinion is divided on whether engineering is an art or a science. Instinctively engineers seek to describe the world exactly but nature confounds them. In response, they find that an intuitive, semi-empirical approach can function better. Hence this is not a "cook-book" and *the answer* is not on page 42. However, it is hoped that the book contains sufficient information to guide engineers through their task to an appropriate solution. Given the limitations of space, where necessary this book refers to standard texts or other existing publications.

Disclaimer

While every effort has been made to check the integrity and quality of the contents, no liability is accepted by either the publisher or the author for any damages incurred as the result of the application of the information contained in this book. Where values for parameters have been stated, these should be treated as indicative only. Readers should independently verify the properties of the sprayed concrete that they are using as it may differ substantially from the mixes referred to in this book.

This publication presents material of a broad scope and applicability. Despite stringent efforts by all concerned in the publishing process, some typographical or editorial errors may occur, and readers are encouraged to bring these to our attention where they represent errors of substance. The publisher and author disclaim any liability, in whole or in part, arising from the information contained in this publication. The reader is urged to consult with an appropriate licensed professional prior to taking any action or making any interpretation that is within the realm of a licensed professional practice.

Acknowledgements

The author would like to acknowledge that the following people and organisations who assisted and/or kindly granted permission for certain figures and photographs to be reproduced in this book:

- Prof Omer Aydan
- Dr Nick Barton
- Dr Stefan Bernard
- Prof Wai-Fah Chen
- Edvard Dahl
- Ross Dimmock
- Dr Yining Ding
- Dr Johann Golser
- Dr Eystein Grimstad
- Esther Casson (nee Halahan)
- Dr Christian Hellmich
- Dr Matous Hilar
- Dr Benoit Jones
- Franz Klinar
- Prof Wolfgang Kusterle
- Mike Murray
- Dr David Powell
- Dr Sezaki
- Ernando Saraiva
- Hing-Yang Wong
- ACI
- BAA plc
- BASF
- D2 Consult
- Landsvirkjun & Johann Kröyer
- Normet
- Morgan Est
- Mott MacDonald Ltd
- J. Ross Publishing, Inc.

- SZDC
- Thameswater
- Tunnelling Association of Canada

The author would also like to express his deep gratitude to all the people that he has worked with and learnt from. Particular thanks are owed to Prof Chris Clayton for his guidance during the original research and to colleagues in the Tunnels Division of Mott MacDonald, especially to Dr David Powell for his inspirational vision and support in the author's career and to Tony Rock for attempting to instill in the author a sense of engineering rigour and lateral thinking.

Author

Alun Thomas is an independent tunnel designer and the founder of All2plan Consulting ApS. Having graduated from Cambridge University in 1994, he went on to work for several major design companies with a short stint working on site for contractors. Alun has a broad experience of many types of tunnelling methods from immersed tubes to segmental linings, from closed face TBMs to hand excavation under compressed air. He is a recognised expert in sprayed concrete lined (SCL/NATM) tunnels and numerical modelling, having completed his PhD at Southampton University in those fields. Alun has been involved in promoting the use of innovative technologies such as permanent sprayed concrete, fibre reinforcement, GFRP rock bolts and spray applied waterproof membranes. He has been involved in many of the recent major UK tunnelling projects such as the Jubilee Line Extension, Heathrow Express, Thameswater Ring Main, Terminal 5, Victoria Station Upgrade and the Elizabeth Line (Crossrail), as well as working on design and construction of tunnels internationally. Alun has been an active member of several industry bodies, notably participating in writing guidelines for BTS, ITA and ITAtech technical committees.

Abbreviations

LIST OF SYMBOLS AND ABBREVIATIONS

α	utilisation factor = stress / strength or deviatoric stress / yield strength
acc.	according to
agg.	aggregate
ACI	American Concrete Institute
B	relaxation time in Kelvin creep model – see 5.6.2
BIM	Building Information Modelling
BTS	British Tunnelling Society
bwc	by weight of cement, other binders and microsilica
ξ	degree of hydration, except in equation 4.1 where it is the "skin factor"
c/c	centre to centre
CCM	convergence confinement method
C/D	cover (depth from ground surface to tunnel axis)/ tunnel diameter
CSL	composite shell lining
Cu	undrained shear strength
D	tunnel diameter
DCA	Degree of composite action
DSL	double shell lining
δ_v	vertical deformation
ε	strain
ε_{dev}	deviatoric strain
E	Young's modulus of elasticity
Ea/R	activation energy = 4000 K – see 5.7
E_{dyn}	dynamic elastic modulus
E_{max}	maximum value of the elastic modulus
E_0	initial tangent modulus
E_{tan}	tangent elastic modulus
eCO2	embodied carbon dioxide content
e_{ij}	deviatoric strain

$\dot{e}_{ij} = \sqrt{2 \cdot \dot{j}_2'}$	deviatoric strain rate
est.	estimated
fc	strength – fcu or fcyl
$f_{fck,fl}$	characteristic peak flexural tensile strength
fcu	uniaxial compressive cube strength
fcyl	uniaxial compressive strength (from tests on cylinders)
FLAC	FLAC3D & FLAC (2D) finite difference program by Itasca
FOB	full overburden pressure
FRS	fibre reinforced shotcrete (sprayed concrete)
f_{R1}	residual flexural strength at crack mouth opening displacement of 0.5 mm (fib 2010)
f_{R3}	residual flexural strength at crack mouth opening displacement of 2.5 mm (fib 2010)
G	elastic shear modulus
G_{vh}	independent shear modulus
GFRP	glass fibre reinforced plastic/polymer
GGBS	granulated ground blast furnace slag
γ	density
γ_f	partial factor of safety for loads (see BS8110)
γ_m	partial factor of safety for materials (see BS8110)
h	depth below groundwater level
HEX	Heathrow Express project
HME	hypothetical modulus of elasticity
HSE	UK Health & Safety Executive
ICE	UK Institution of Civil Engineers
ITA	International Tunnelling Association
JLE	Jubilee Line Extension project
J_2	second deviatoric invariant of principal stresses
J_2	$(1/6) \cdot ((\sigma_\sigma - \sigma_\sigma)^2 + (\sigma_\sigma - \sigma_\sigma)^2 + (\sigma_\sigma - \sigma_\sigma)^2)$
k	permeability
K	elastic bulk modulus
K_0	ratio of horizontal effective stress to vertical effective stress
long.	longitudinal
λ	stress relaxation factor
max.	maximum
η	viscosity, except in equation 2.1 where it is a constant
NATM	New Austrian Tunnelling Method
OCR	overconsolidation ratio
OPC	ordinary Portland cement
p	mean total stress
p'	mean effective stress
PCL	partial composite lining
PFA	pulverised fly ash

PSCL	permanent sprayed concrete lining
pts	points
Q	water flow
r	tunnel radius (e.g. equation 4.1)
r	deviatoric stress $= 2.(J_2)^{0.5}$
R	tunnel radius
R	universal constant for ideal gas – see Ea/R above
RH	relative humidity
σ	stress (compression is taken to be negative)
σ_v	vertical stress
$\sigma_1, \sigma_2, \sigma_3$	principal stresses
SAWM	spray applied waterproofing membrane
SCL	sprayed concrete lined/lining
SFRS	steel fibre reinforced shotcrete (sprayed concrete)
SSL	single shell lining
str	strength

θ Lode angle (or angle of similarity) where $\cos\theta = \dfrac{2\sigma_1 - \sigma_2 - \sigma_3}{2\sqrt{3J_2}}$

corresponds to the tensile meridian and $\theta = 60°$ corresponds to the compressive meridian

t	time or age, except in equations 4.1 and 6.1 where is denotes thickness
TSL	thin spray-on liner or thin structural liner
v, υ	Poisson's ratio
V	ultrasonic longitudinal wave velocity
w/c	water/cement ratio
w.r.t.	with respect to
2D	two dimensional
3D	three dimensional

SUBSCRIPTS

A_1	in the first principal stress/strain direction
A_c	compressive
A_g	related to the ground
A_h	in the horizontal plane/direction
A_{hh}	in the horizontal plane/direction
A_{ij}	in principal stress/strain directions where i and j can be 1,2 or 3
A_o	initial value (e.g.: value of modulus at strain is zero)
A_t	time-dependent value
A_{tan}	tangential (e.g.: tangential elastic modulus)
A_u	undrained (in context of geotechnical parameters)

A_v in the vertical plane/direction
A_{xx} in the direction of x-axis
A_{yy} in the direction of y-axis
A_{zz} in the direction of z-axis
A_{28} value at an age of 28 day

Chapter 1

What is an SCL tunnel?

Considering the title of this book, this seems like the first question that should be answered. An SCL tunnel is a tunnel with a sprayed concrete lining. This generic definition makes no claims on how the tunnel was designed, the ground it was built in or what its purpose is. It simply describes the type of lining used.

Modern SCL tunnel construction is described in more detail in Section 1.3. Figure 1.1 and Figure 1.2 show a typical excavation sequence and cross-section for a large diameter tunnel in soft ground at a shallow depth. The arrangement of the excavation sequence is influenced by the geometry of the tunnel, the stability of the ground and the construction equipment. In shallow tunnels, it is important to close the invert as close to the face as possible, in order to limit ground deformations. However, the designer has a fair degree of freedom in choosing the exact arrangement of the excavation sequence.

After each stage of the excavation sequence has been mucked out, sprayed concrete is sprayed on the exposed ground surface. The lining is often built up in several layers with mesh reinforcement inserted between the layers. Alternatively, short fibres can be added to the mix to provide some tensile capacity. Once that section of lining is complete, the next stage is excavated, and so the process progresses and a closed tunnel lining is formed. Often, the sprayed concrete lining does not form part of the permanent works and another lining is installed at a later date (see Figure 1.1).

In rock tunnels, sprayed concrete works in concert with rock bolts to support the rock (see Figure 1.3). As such, the sprayed concrete is an important part of the support and often forms part of the permanent support (Grov 2011). As with tunnels in soft ground, the degree and timing of support and the excavation sequence are governed by the stability of the ground.

A plethora of other terms exist for tunnels with sprayed concrete linings: most famously in Europe, there is NATM – the New Austrian Tunnelling Method; in North America, SEM – Sequential Excavation Method – is often used; while elsewhere no particular emphasis is placed on the use of sprayed concrete as a distinguishing feature, for example, in hard rock tunnelling. In this book, the term SCL will be used throughout as a descriptive

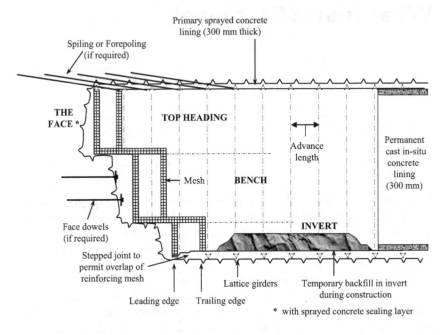

Figure 1.1 Long-section of an SCL tunnel in soft ground.

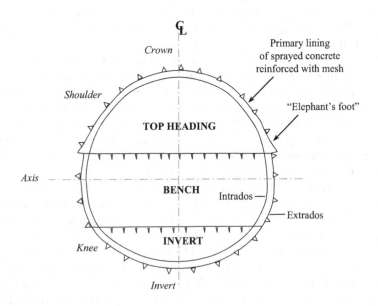

Figure 1.2 Cross-section of an SCL tunnel in soft ground.

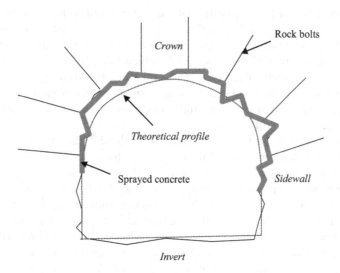

Figure 1.3 Cross-section of an SCL tunnel in rock.

term, and as such it opens the door to look at the many and varied uses of sprayed concrete in modern tunnelling.

1.1 SPRAYED CONCRETE – THE EARLY DAYS

The invention of sprayed concrete is generally attributed to Carl Ethan Akeley in 1907, who used a dry mix sprayed mortar to apply a durable coating to dinosaur bones. However, in Germany, August Wolfsholz had been developing equipment for spraying cementitious mortar in tunnels for rock support from as early as 1892 (Strubreiter 1998), and Carl Weber patented a method for spraying concrete in 1919 (Atzwanger 1999). While sprayed concrete was used on a few engineering projects to repair concrete structures or for rock support in the first half of that century – it was even trialled by the Modernist architect Le Corbusier for one of his projects – this material and method first attracted serious attention after its use on a series of pioneering projects in Venezuela and Austria by Ladislaus von Rabcewicz in the 1950s (Rabcewicz 1969). Sprayed concrete and mortars can have a multitude of uses including architectural purposes, fire protection and even 3D printing. This book focuses on their uses in tunnelling only.

Early sprayed concrete was not a high-quality product. Large quantities of aggressive accelerating additives were required to get the sprayed concrete to adhere to the ground and so that reasonably thick layers could be sprayed. The environment during spraying was very unhealthy due to the large quantities of dust and the caustic nature of the accelerators. Despite the accelerators, a large quantity of the sprayed concrete failed to adhere and fell as waste material onto the tunnel floor – so-called "rebound". The material was very sensitive to the influence of the nozzleman since he controlled how the material was sprayed (which determines the compaction) and the water content. Because of this and a deterioration caused by accelerators such as "water-glass", the long-term strength of sprayed concrete was much lower than conventionally cast concrete and the material was more variable in quality.

Hence, research and development since the 1970s have focused primarily on accelerators and admixtures (to achieve higher early strengths with lower dosages of these expensive additives, without compromising long-term strength and to reduce dust and rebound) and spraying equipment (to improve the quality, spraying quantity and automation). Research into the durability and mechanical properties of sprayed concrete other than strength and stiffness followed later as the early challenges were overcome and the design approaches and usage developed. There is an increasing trend nowadays towards the use of permanent sprayed concrete linings.

1.2 WHY USE SPRAYED CONCRETE LININGS?

To understand the origin and merits of sprayed concrete tunnel linings, one must first appreciate some fundamental tunnelling principles.

1. Tunnelling is a case of three-dimensional soil–structure interaction.
2. The load to be carried by the composite structure of the ground and lining arises from the *in-situ* stresses and groundwater pressure.
3. Deformation of the ground is inevitable, and it must be controlled to permit a new state of equilibrium to be reached safely.
4. The unsupported ground has a finite "stand-up time".
5. Often the strength of the ground depends on how much it is deformed.
6. The load on the lining will depend on how much deformation is permitted and how much stress redistribution ("arching") within the ground is possible.
7. The art of tunnelling is to maintain as far as possible the inherent strength of the ground so that the amount of load carried by the structure is minimised.

These basic principles have been understood implicitly or explicitly by experienced tunnellers since tunnels were first constructed. However, they were brought to the forefront of attention by the pioneering work of engineers, such as Rabcewicz, who developed the tunnelling philosophy that is now marketed as the NATM. In his early work in rock tunnels, Rabcewicz (1969) recognised that sprayed concrete was a material well suited to tunnelling for the reasons below.

- Sprayed concrete is a structural material that can be used as a permanent lining.
- The material behaviour of sprayed concrete (which is initially soft and creeps under load but can withstand large strains at an early age) is compatible with the goal of a lining which permits ground deformation (and therefore stress redistribution in the ground).
- The material behaviour (specifically the increase in stiffness and strength with age) is also compatible with the need to control this deformation so that strain softening in the ground does not lead to failure.
- Sprayed concrete linings can be formed as and when required and in whatever shape is required. Hence the geometry of the tunnel and timing of placement of the lining can be tailored to suit a wide range of ground conditions. Sprayed concrete can also be combined with other forms of support such as rock bolts and steel arches.

One may also note the lower mobilisation times and costs for the major plant items compared to tunnel boring machines (TBMs). The same equipment can be used for shaft construction as well as tunnelling. SCL tunnelling offers a freedom of form that permits tunnels of varying cross-sections and sizes and junctions to be built more quickly and cost-effectively than if traditional methods are used.

1.3 DEVELOPMENT OF SCL TUNNELLING

Sprayed concrete was first used as temporary (and permanent) support in rock tunnels. However, the principles above apply equally to weak rocks and soils. In the 1970s shallow SCL tunnels were successfully constructed in soft ground as part of metro projects in cities such as Frankfurt and Munich.

Taking the UK as one example, Figure 1.4 charts the rise of SCL tunnelling. This technique arrived relatively recently to the UK and has only become widely used within the last 25 years. Initially there was great enthusiasm for SCL tunnelling. However, following the collapse of a series of SCL tunnels in 1994, this construction method came under intense scrutiny. Vociferous sceptics asserted at the time that SCL tunnelling cannot and should not be used in soft ground at shallow depths (e.g. Kovari (1994)).

Reports by the UK Health and Safety Executive (HSE 1996) and the Institution of Civil Engineers (ICE 1996) have established that SCL tunnels can be constructed safely in soft ground, and the reports provided guidance on how to ensure this during design and construction. The reports also drew attention to the weaknesses of this method:

- The person spraying the concrete (the nozzleman) has a considerable influence over the quality of the lining so the method is vulnerable to poor workmanship. This is particularly true for certain geometries of linings.
- The performance of the linings and ground must be monitored during construction to verify that both are behaving as envisaged in the design. The data from this monitoring must be reviewed regularly in a robust process of construction management to ensure that abnormal behaviour is identified and adequate countermeasures are taken.
- It is difficult to install instrumentation in sprayed concrete linings and to interpret the results (Golser et al. 1989, Mair 1998, Clayton et al. 2002).
- It is difficult to predict the behaviour of SCL tunnels in advance.

The specific disadvantages of this method, as applied to soft ground, are:

- Minimising deformations is of critical importance. Otherwise strain-softening and plastic yielding in the ground can lead rapidly to collapse. Complex excavation sequences can lead to a delay in closing the invert of the tunnel (and forming a closed ring). This delay can permit excessive deformations to occur.
- In shallow tunnels, the time between the onset of failure and total collapse of a tunnel can be very short, so much tighter control is required during construction.

More general disadvantages include the fact that advance rates are slower than for TBM-driven tunnels so SCL tunnels are not economic for long tunnels with a constant cross-section (i.e. greater than about 500 m to a few kilometres, depending on ground conditions). A higher level of testing is required for quality control during construction, compared to a segmentally lined tunnel.

Following the HSE and ICE reports, a considerable amount of guidance has been produced on such subjects as certification of nozzlemen (Austin et al. 2000, Lehto & Harbron 2011), instrumentation and monitoring (HSE 1996) and risk management (BTS/ABI 2003). EFNARC now operates a nozzleman certification scheme which is endorsed by the ITA. The UK tunnelling industry has incorporated much of this into its standard practices. Since the collapse at Heathrow, more than 500,000 m^3 of shallow SCL tunnels have been successfully constructed in a variety of soft ground conditions. Major UK projects such as the Heathrow Baggage Transfer Tunnel (Grose & Eddie 1996), the CTRL North Downs Tunnel (Watson et al. 1999) and Crossrail (Smith 2016) have demonstrated the great benefits to be gained from this method, not least in terms of time and cost savings compared to traditional construction methods.

So, the reputation of sprayed concrete in UK tunnelling has recovered, and for certain types of work SCL has supplanted traditional methods. For

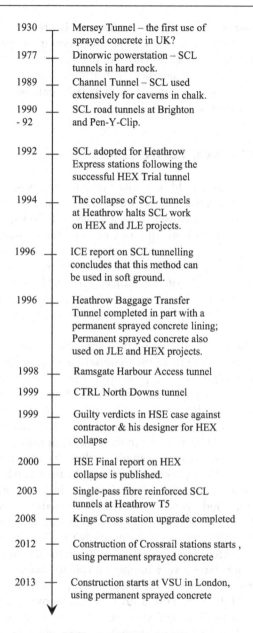

1930	Mersey Tunnel – the first use of sprayed concrete in UK?
1977	Dinorwic powerstation – SCL tunnels in hard rock.
1989	Channel Tunnel – SCL used extensively for caverns in chalk.
1990 - 92	SCL road tunnels at Brighton and Pen-Y-Clip.
1992	SCL adopted for Heathrow Express stations following the successful HEX Trial tunnel
1994	The collapse of SCL tunnels at Heathrow halts SCL work on HEX and JLE projects.
1996	ICE report on SCL tunnelling concludes that this method can be used in soft ground.
1996	Heathrow Baggage Transfer Tunnel completed in part with a permanent sprayed concrete lining; Permanent sprayed concrete also used on JLE and HEX projects.
1998	Ramsgate Harbour Access tunnel
1999	CTRL North Downs tunnel
1999	Guilty verdicts in HSE case against contractor & his designer for HEX collapse
2000	HSE Final report on HEX collapse is published.
2003	Single-pass fibre reinforced SCL tunnels at Heathrow T5
2008	Kings Cross station upgrade completed
2012	Construction of Crossrail stations starts , using permanent sprayed concrete
2013	Construction starts at VSU in London, using permanent sprayed concrete

Figure 1.4 The development of SCL tunnelling in the UK.

example, SCL is the method of choice for shafts and short tunnels in the London Clay, and SCL was the obvious choice for the five mined stations on the Crossrail project in London. Even more significantly, Crossrail adopted sprayed concrete as the permanent lining for most parts of those stations (Dimmock 2011). This pattern of development mirrors the experience of many other countries. Nevertheless, SCL tunnels are still perceived to be difficult to design because of the complex behaviour of sprayed concrete. This uncertainty, coupled with a history of high-profile failures, means that SCL tunnelling is perceived as risky. In truth, SCL tunnelling is no more risky than any other type of tunnelling. The risks can be identified clearly and successfully managed.

1.4 SAFETY AND SCL TUNNELLING

Safety is a subject that should be treated holistically. Whether one considers a near miss, a long-term illness or a tunnel collapse with fatalities, one usually finds that the root causes lie in a series of failings, and often steps could have been taken by both the designer and the construction team to avoid the incident. The modern approach to safety emphasises that everyone involved in the project has a role in ensuring the safety of themselves and others. The whole life of the project – right through to decommissioning and demolition – should be considered as well as recognising that occupational health is as important as immediate safety and environmental protection.

SCL tunnels should be treated in the same way as any other construction activity. Hazards arise from the materials used (e.g. chemicals used as accelerators), the application (e.g. projecting a jet of concrete at high velocity or the overall stability of a tunnel) and the finished product (e.g. steel fibres protruding from the surface). The impacts of the hazards are best understood by means of a risk assessment. The hazards stemming from sprayed concrete should be assessed in the context of the other construction activities and their hazards since sometimes these hazards interact. General advice on safety including sprayed concrete application can be found in publications from the ITA (e.g. WG5 Safe Working in Tunnelling, ITA 2004, and Guidelines for good occupational health and safety practice in tunnel construction, ITA 2008). Readers should also ensure that they understand how the relevant local safety regulations apply to SCL tunnelling.

As noted above, safety is not merely a question for the construction team to address. Designers have a very important role to play as do clients. Risk-based design is discussed in Section 4.1.2. The reports on the Heathrow collapse (ICE 1996, HSE 1996, HSE 2000) are instructive reading for anyone involved in SCL tunnelling, and they explore how failings in a multitude of areas combined to lead to that particular accident. Since most hazards emerge in the construction phase, safety is examined in detail in passages on construction management (see Section 7.1).

Chapter 2

Sprayed concrete

2.1 CONSTITUENTS AND MIX DESIGN

"Sprayed concrete is concrete which is conveyed under pressure through a pneumatic hose or pipe and projected into place at high velocity, with simultaneous compaction" (DIN 18551 1992). It behaves in the same general manner as concrete but the methods of construction of SCL tunnels and of placement of sprayed concrete require a different composition of the concrete and impart different characteristics to the material, compared to conventionally placed concrete. Sprayed concrete consists of water, cement and aggregate, together with various additives. On a point of nomenclature, sprayed concrete is also known as "shotcrete", while "gunite" normally refers to sprayed mortar, i.e. a mix with fine aggregates or sand only.

The composition of the concrete is tailored so that:

- It can be conveyed to the nozzle and sprayed with a minimum of effort.
- It will adhere to the excavated surface, support its own weight and the ground loading as it develops.
- It attains the strength and durability requirements for its purpose in the medium to long term.

Table 2.1 contains a comparison of the constituents of a high-quality sprayed concrete and an equivalent strength cast *in-situ* concrete. Considering each component in turn, one may note that:

- The water–cement ratio in sprayed concrete is higher so that the mix can be pumped and sprayed easily.
- Ordinary Portland cement is normally used, in conjunction with cement replacements such as pulverised fly ash (PFA), though special cements are sometimes used.
- The mix is "over-sanded" to improve pumpability (Norris 1999) (see Figure 2.1 for grading curve).
- The maximum aggregate size is usually limited to 10 or 12 mm.

Table 2.1 Typical mix design

	High-quality wet mix sprayed concrete (Darby & Leggett 1997)	Cast in-situ concrete (from Neville 1995)
Grade	C40	C40
Water–cement ratio	0.43	0.40
Cement inc. PFA, etc.	430 kg/m³	375 kg/m³
Accelerator	4–8%	–
Plasticiser	1.6% bwc	1.5%
Stabiliser	0.7% bwc	–
Microsilica	60 kg/m³	–
Max. aggregate size	10 mm	30 mm
Aggregate < 0.6 mm	30–55%	32%

- Additives are used to accelerate the hydration reaction (see Figure 2.2 for the effect of increasing accelerator dosage on strength gain).
- Plasticisers and stabilisers are added to improve workability as in conventional concrete.
- Other components may include micro silica, which is added to improve immediate adhesion (thereby allowing the accelerator dosage to be reduced) and to improve long-term density (which improves strength and durability) or fibres, which are added for structural reinforcement or crack control.

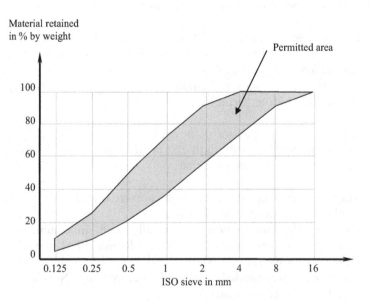

Figure 2.1 Typical grading curve for sprayed concrete (BTS 2010).

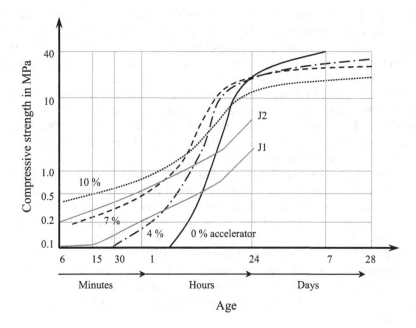

Figure 2.2 Early age strength gain depending on dosage of accelerator with ÖBV J-curves for minimum strength (after Kusterle 1992).

Each component represents a large subject in itself, and more information on most of them can be found in standard textbooks on concrete technology (e.g. Byfors 1980, Neville 1995). Where relevant, their influence on the mechanical properties of sprayed concrete will be briefly discussed in Section 2.2. More details on the constituents of sprayed concrete can be found in the International Tunnelling Association's state of the art review (ITA 1993, 2010) and other texts (e.g. Austin & Robins 1995, Brite Euram 1998, Brooks 1999, Melbye 2005, ACI 506.2 2013, ACI 506R-16 2016). The mix designs in certain regions have developed particular characteristics as a result of the ingredients available.

Finally, it is worth noting that sprayed concrete technology remains a fluid field with fits and starts of innovation. We can look forward to new materials being applied here, driven by the impulse to improve the efficiency in this demanding application. As an example, nanotechnology has introduced the possibility of using graphene oxide in sprayed concrete (Papanikolaou et al. 2018).

2.1.1 Cement

For wet mix sprayed concrete, ordinary Portland cement (OPC) is normally used – alone or blended with cement replacements, i.e. CEM I, II or III. The accelerator is used to speed up the hydration so a "rapid hardening cement" is not needed. However, the chemistry of the cement is very important as well as its compatibility with the accelerator. For example, if the percentage of the fast-setting component, tricalcium aluminate, is unusually low, the cement may react too slowly for use in sprayed concrete. Careful mix design and pre-construction trials are essential (see Section 7.1.1).

Dry mix sprayed concrete also normally uses OPC. However, to reduce the need for accelerators, new types of cement – so-called "spray cements" – have been developed for use with the dry mix process (Testor 1997, Lukas et al. 1998). If gypsum (hydrated calcium sulphate) is removed from cement, the speed of the hydration reaction increases dramatically. Normally the gypsum reacts to form a film of calcium sulphoaluminate (ettringite) on the surface of the tricalcium aluminate in cement particles. Otherwise, the tricalcium aluminate is free to react immediately and form hydrated calcium aluminate directly (Neville 1995, Atzwanger 1999). The reaction is so rapid that most of these new cements can only be used with oven-dried aggregate; otherwise hydration may occur in the delivery hoses. No accelerator is required. The latest "spray cements" can also be used with naturally moist aggregates. While costs are reduced by not having to use accelerators, extra costs are incurred in the preparation and storage of the cement and aggregate.

Alternatively, calcium sulphoaluminate (CSA) cements can be used for extremely rapid hardening concrete (e.g. Mills et al. 2018), typically as in the dry mix format. While the carbon footprint of these products is more attractive, development is still required to create economically competitive mixes.

2.1.2 Cement replacements

Pulverised fly ash (PFA) and ground granulated blast furnace slag (GGBS) are added to the sprayed concrete mix as cement replacements in the normal manner, though GGBS cannot be used in the same quantities as in conventional concrete. Because of its particles' angular shape GGBS can only be used to replace up to 35% of the cement (Brite Euram 1998). Above this level, there are problems pumping the mix. Furthermore, GGBS is relatively slow reacting. Metakaolin is another possible cement replacement under research these days (Bezard & Otten 2018). Since these materials react more slowly than cement, their beneficial contributions to durability characteristics (e.g. Yun et al. 2014), density and strength are only seen over the longer term (i.e. at ages greater than 28 days). Cement replacements can also help to reduce sintering (Thumann et al. 2014, Bezard & Otten 2018) because there is less free lime (see Section 4.2.3). Generally, they seem to be more effective (and in particular metakaolin and microsilica) than one would expect based on the percentage of cement replaced (Thumann et al. 2014). These cement replacements and geopolymers (alkali-activated alumino-silicates) more generally are seen as a way to reduce the carbon footprint of sprayed concrete – see Section 2.1.8. There is some interest in developing accelerators which can target these relatively slow-reacting binders (e.g. Rudberg & Beck 2014, Myrdal & Tong 2018).

One issue for all cement replacements which come from industrial by-products is the availability of supply. Not only are these binders not universally available, but also, they may even become more scarce, if they originate from environmentally unfavourable processes such as coal combustion.

2.1.3 Water

Ordinary water is used for sprayed concrete. As in conventional concrete, the water–cement ratio has a large influence on the strength of the concrete. Ideally, the water–cement ratio should be less than 0.45. In the wet mix process, the water is added during batching as in normal concrete. In the dry mix process, the water is added in the tunnel during spraying. The wet and dry mix processes are described in more detail in Section 3.4.

2.1.4 Sand and aggregate

Sand and aggregate form the bulk of sprayed concrete. The normal rules for concrete govern the choice of rock for the aggregate. The pumping and spraying process places onerous demands on the mix. A smooth grading curve is essential, and rounded aggregates are preferred to angular particles. As noted above, there tend to be more fine particles in a sprayed concrete mix. Contamination of the mix with over-sized stones (i.e. larger than about 10 mm in diameter) is a common cause of blockages during spraying. Blockages can cause costly delays and wastage of concrete. Therefore, careful design of the mix and control of the batching is advisable.

Moisture in the aggregate contributes to the water in the mix, and this affects properties such as the strength. Sometimes special measures are needed to control the moisture content.

2.1.5 Accelerators

Almost all sprayed concrete mixes require an accelerator to speed up the hydration in order to achieve the early age strength that is needed (Myrdal 2011) – see also section 2.1.1. The accelerators were one of the main problems with sprayed concrete in the early days. Although they could achieve high early strengths, the chemicals – such as those based on aluminates – were very caustic and so posed a danger to the workers. Sometimes also the products of hydration were unstable, and the strength of the sprayed concrete actually decreased over time, notably when using water-glass (modified sodium silicate).

Table 2.2 Acceptable setting times for accelerated cements (Melbye 2005)

	Caustic accelerators	Alkali-free accelerators
Initial set	< 60 seconds	< 300 seconds
Final set	< 240 seconds	< 600 seconds

Modern accelerators do not have these problems. They are normally based on combinations of aluminium salts (sulphates, hydroxides and hydroxysulphates) (DiNoia & Sandberg 2004, Myrdal 2011). The modern accelerators are classed as "non-caustic" so they are safer to use. They are also "alkali free" – equivalent Na_2O content $< 1.0\%$ – which reduces the risk of alkali–silica reaction in the concrete. In wet mix, the accelerator is added in liquid form at the nozzle during spraying. Dry mix uses the same approach, but the accelerator can also be added as a fixed dosage in powder form when using pre-bagged mixes. The only drawback of the modern accelerators is that they do not act as fast as the old caustic ones – see Table 2.2. The typical dosage for accelerators ranges from 5 to 10% by weight of cement.

Some products on the market today are gelling agents rather than chemicals which accelerate the hydration process. The two products should not be confused. Although a gelling agent will help the concrete to adhere to the substrate, thick layers cannot be sprayed using a gelling agent because it is insufficiently strong to hold the self-weight of the concrete. This also means that, until the concrete starts to hydrate, it will not be able to carry any load from the ground. Since this is a specialist field it is best to consult with accelerator manufacturers on how best to use their products. Laboratory and/or field tests are needed to check that the performance of the accelerators.

As an aside, foam concrete (also known as cellular concrete) can be sprayed without an accelerator. The compaction during spraying drives out the air in the foam concrete, creating sufficient cohesion for the concrete to remain in place but, because it has not been accelerated, it remains soft enough to the carved and finished. Yun et al. (2018) describe an innovative use of this technique to create sculptures at tunnel portals.

2.1.6 Admixtures

To meet the conflicting demands of the design strengths (both short- and long-term), a long pot life, ease of pumpability and sprayability, a cocktail of admixtures is added.

Plasticisers (lignosulphates) and superplasticisers (naphthalenes/melamines or modified polycarboxylic esters) increase workability without increasing the water–cement ratio (except for the water contained in the plasticiser itself).

Retarders slow down hydration and thereby extend the pot life of the concrete. For sprayed concrete, if retarders are used with a set accelerator, the early age strength development will also be slowed down, making it difficult to achieve acceptable early age strengths in practice. Activators can be used to remove the inhibiting effects of retarders. Alternatively, "stabilisers" or "hydration control agents" work differently to retarders in that they prevent hydration. These types of admixtures can prevent the start of the hydration process for significantly longer than retarders, and, when combined with set accelerators, they do not reduce the early age strength development of the sprayed concrete. Some manufacturers claim that their products can extend the pot life of a wet mix from the normal 1.5 hours to as much as 72 hours (Melbye 2005).

The interaction of the admixtures depends on the exact mix recipe, and sometimes combinations can produce unexpectedly adverse results. For example, Niederegger and Thomaseth (2006) discuss the effect of certain plasticisers causing excessive stickiness which in turn can lead to variable early age strengths. Since this is a specialist field it is best to consult with admixture manufacturers on how best to use their products. Laboratory and/or field tests are needed to check the performance of the admixtures.

2.1.7 Microsilica

Microsilica has many benefits and its usage is discussed in Sections 2.2.1 and 2.2.9.

2.1.8 Mix design

The subject of mix design for concretes is discussed in detail in Neville (1995). The main difference for sprayed concrete is the addition of early age strength and sprayability criteria. That is not to say that the process of selecting a mix to meet the design and construction criteria is an easy one. Minor variations in one component can have a large impact on the performance of the mix overall. So, although in practice engineers often rely on an empirical approach, using tried and tested mix designs, it is naïve to believe that a recipe which worked on one project will be as good on another project, using different cement, aggregate or other ingredients. The development of a mix design is guided by the results of laboratory and field tests. It is important to allow sufficient time for this sort of pre-construction testing so that a good mix is available before tunnelling begins (see also Section 7.2.1).

2.1.9 Environmental sustainability

The concrete lining represents one of the largest contributors to the environmental impact of a tunnel. The high cement content in sprayed concrete exacerbates this as do the additional admixtures needed. On the other hand,

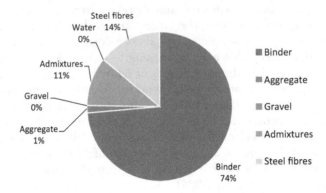

Figure 2.3 Components of embodied carbon in a sprayed concrete lining.

sprayed concrete linings tend to have lower contents of reinforcement than other reinforced concrete structures, especially if fibres are used. Figure 2.3 shows an indication of the relative importance of each component in terms of embodied carbon content (eCO_2).

Of course, the embodied carbon is only one measure of our environmental impact. A life cycle assessment offers a more holistic evaluation of environmental impact (Kodymova et al. 2017). However, eCO_2 is relatively easy to estimate, and an increasing number of projects are calculating their carbon footprints. Having estimated the impact, the next logical step is to seek to reduce it.

A variety of steps can be taken to do this:

- Minimising the transport distances
- Use of recycled aggregates
- A higher proportion of recycled steel
- Cement replacements (see Section 2.1.2)
- Permanent sprayed concrete designs which tend to be thinner than the traditional "double shell" approach (i.e. avoiding the wasteful use of temporary linings) – see Section 4.2.4
- Macrosynthetic fibres instead of steel fibres (if appropriate)

As an aside, if macrosynthetic fibres are used, the rebound and muck from a tunnel need to be screened to avoid contamination of the ground or water courses with these plastic fibres (Crehan et al. 2018).

Table 2.3 lists the main parameters for a comparison of the carbon footprint for various design approaches. This is based on a large diameter (11.5 m) tunnel in hard rock, with a drained design concept – i.e. water pressure is relieved by drains in the invert, but there is a waterproofing layer above the invert. This comparison covers the lining only since the excavation method

Table 2.3 Key parameters of eCO$_2$ calculation

Item	Hard rock PSCL (steel fibre)	Hard rock PSCL (macrosynthetic)	Hard rock DSL
Primary lining concrete thickness	80	80	80
Primary fibre/bar	40	8	40
Reinforcement type	Steel fibre	Macrosynthetic fibre	Steel fibre
Regulating layer	40	40	0
Membrane type	SAWM	SAWM	PVC sheet
Secondary lining concrete thickness	80	80	300
Secondary steel fibre/bar	40	8	97
Reinforcement type	Steel fibre	Macrosynthetic fibre	Steel bar

is the same for all options. No allowance has been made for the possibility of reducing the size of the tunnel in the PSCL options.

Figure 2.4 shows that the PSCL option could offer a saving of more than 50% in terms of the carbon footprint of the lining. In this context, the benefit of the sprayed applied waterproofing membrane (SAWM) lies not in its own, slightly lower embodied carbon content per tonne (compared to a sheet membrane and geotextile) but rather the fact that its bonded nature permits the secondary lining to be sprayed onto it. The savings in concrete and steel quantities are significant. The PSCL options have a lower embodied carbon content than the traditional double shell lining (DSL), despite

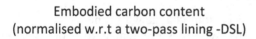

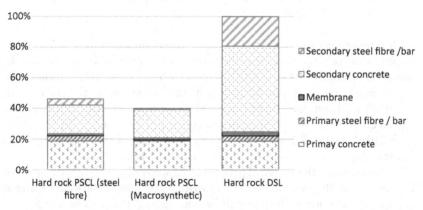

Figure 2.4 Embodied CO$_2$ for different design options for a hard rock tunnel (normalised w.r.t. a two-pass lining – DSL).

the fact that sprayed concrete has a higher embodied carbon content per m^3 than cast concrete – as noted above. Similarly, steel fibres have a higher embodied carbon content than plain reinforcing steel bars. However, the weight of fibres is much lower in a PSCL secondary than the weight of bar reinforcement in a cast in place secondary lining.

The difference in a similar comparison for weak rock and soft ground cases is less pronounced since the secondary sprayed linings tend to be thicker. Nevertheless, in those cases, the PSCL design approach can offer savings of around 25 and 10% respectively. Obviously, the values in such comparisons will vary, depending on the specific characteristics of each particular project.

Finally, another environmental challenge, which SCL tunnels will face in the future, will be the change in availability of aggregates. Naturally occurring deposits of aggregates with grading curves suitable for spraying will become increasingly scarce, and mixes will have to use blended aggregates (e.g. Yun et al. 2011). This will increase the eCO_2 for the aggregates as well as their cost. However, the former will have a negligible impact on the overall carbon footprint.

2.2 MATERIAL PROPERTIES AND BEHAVIOUR

Considering sprayed concrete as a construction material, we could start by asking a series of basic questions.

- How strong is sprayed concrete?
- Is it brittle or ductile?
- Do its properties or behaviour change with time?
- Do its properties change with pressure, temperature or other environmental conditions?

The following sections will try to answer these questions and more. As an introduction, Table 2.4 contains the properties of a sprayed concrete and an equivalent strength cast concrete mix described in Table 2.1. Although the properties of this sprayed concrete are at the higher end of the typical values for sprayed concrete, there is a trend towards using such higher quality sprayed concrete as the norm on major construction projects (Brooks 1999, Smith 2016). Figure 2.2 shows the strength of sprayed concrete at ages less than one day in comparison to an unaccelerated mix.

The variation in the material properties of sprayed concrete with age will be discussed in the next sections. Following them, the variation of sprayed concrete's behaviour with time will be considered. This can be subdivided into two categories: stress-independent changes (due to shrinkage and temperature effects – Section 2.2.6) and stress-dependent changes (due to

Table 2.4 Typical properties of sprayed and cast concrete

Property	High-quality sprayed concrete	Cast *in-situ* concrete
Compressive strength @ one day in MPa	20	6 (est.)
Compressive strength @ 28 days in MPa	59	44
Elastic modulus @ 28 days in GPa	34	31 (est.)
Poisson's ratio, v, @ 28 days	0.48–0.18 [a]	0.15–0.22
Tensile strength @ 28 days in MPa	> 2 (est.) [b]	3.8 (est.)
Initial setting time (start–end) in mins	3–5 [c]	45–145 (est.)
Shrinkage after 100 days in %	0.1–0.12	0.03–0.08
Specific creep after 160 days in %/MPa	0.01–0.06	0.008
Density kg/m^3	2,140–2,235	2,200–2,600
Total porosity in %	15–20 [d]	15–19
Permeability in m/s	10^{-11} to 10^{-12}	10^{-11} to 10^{-12}
Average water penetration	< 25 mm	< 25 mm
Microcracking @ 28 days in cracks/m	1,300	–
Coefficient of thermal expansion in -/K	$8.25–15 \times 10^{-6}$ [e]	10×10^{-6} [f]
Slump in mm	180–220 [g]	50

[a] Kuwajima 1999.
[b] Kuwajima 1999.
[c] See Table 2.2.
[d] Blasen 1998 and Lukas et al. 1998.
[e] Kuwajima 1999 and Pottler 1990.
[f] Eurocode 2 (2004) Cl. 3.1.3.
[g] BASF 2012.

creep – Section 2.2.7). The variation with material properties with age with respect to durability is covered in Section 2.2.9.

2.2.1 Strength in compression

Theories and mechanisms

Strength is often the first parameter that an engineer examines when considering a new material. As with all materials, the strength of concrete is governed as much by the flaws and imperfections within the material as by the intrinsic strengths of the main components and their interaction. In the case of concrete, the main components are the hydrated cement paste and the aggregate.

Typical compressive strengths of the hydrated cement paste (in the form of very dense cement paste compacts) can be up to 300 to 500 MPa, while the compressive strength of rocks commonly used for aggregate lies between 130 and 280 MPa (Neville 1995). As the concrete hydrates, the strength increases – see Appendix A for formulae to predict the increase in strength with age.[1] The compressive strength at an early age is critical to the safety of a tunnel, and it can be measured in a variety of ways – see Section 7.2.2. Gibson and Bernard (2011) describe a method using ultrasound to measure compressive strength and found that this worked well at strengths from 1 to 10 MPa and within the first 72 hours. The formula is:

$$f_{cyl} = \left(e^{V/500}\right)/80 \qquad (2.1)$$

where f_{cyl} is the uniaxial compressive strength in MPa and V is ultrasonic longitudinal wave velocity.

The imperfections are voids or pores, microcracks and macrocracks (both due to shrinkage and loading). The total porosity of concrete typically ranges between 15 and 20% of the volume (see Table 2.5). The porosity comprises gel pores (between the individual crystals and particles of gel; 10^{-7} to 10^{-9} m in size) and capillary pores (10^{-4} to 10^{-7} m in size), which remain after hydration and are partially occupied by excess water, and air pores (10^{-2} to 10^{-4} m in size), which may be either intentional (entrained air pores) or accidental (due to poor compaction). The porosity of sprayed concrete tends to lie at the higher end of the range for concretes (Kusterle 1992, Lukas et al. 1998, Blasen 1998, Oberdörfer 1996, Myren & Bjontegaard 2014) – see Table 2.5 – with the highest porosities generally in wet mix sprayed concrete. Holter and Geving (2016) found suction porosity (the sum of capillary and gel pores) of wet mix samples to be around 20% which is close but about 1% higher than standard methods would suggest. Myren and Bjontegaard (2014) report suction porosities for FRS between 17 and 19%. Considering a wet mix sprayed concrete, a dry mix sprayed concrete (spray cement with moist aggregate) and a normal cast concrete, all with the same water–cement ratio of 0.55, Blasen (1998) found that porosity of the wet mix was 16% greater than the cast concrete, while the porosity of the dry mix was only 8.7% higher. Consequently, wet mix sprayed concretes may tend to achieve lower strengths than comparable dry mixes.

Failure of concrete in compression is governed by cracking under uniaxial or biaxial compression and by crushing under multi-axial stress (Neville 1995, Chen 1982). Existing microcracks due to hydration and drying shrinkage start to grow when the load exceeds about 30% of the maximum compressive strength of mature concrete (Feenstra & de Borst 1993). These microcracks are mainly located at the interface between the aggregate and hardened cement paste. As the size of the microcracks increases, the effective area resisting the applied load decreases, and so the stress rises locally faster than the nominal load stress (Neville 1995). This leads to strain

Table 2.5 Composition of porosity

Pore type	Pore diameter	Mix type	% of total volume
Gel	< 0.1 μm	Dry and wet	3–4% est. [a]
Capillary	0.1–10 μm	–	15–19% [a]
		Wet	13–17% [b]
		Dry and wet	17.5% [c]
Entrained air pores and accidental voids	> 10 μm and 0.001 to 0.1 m	Wet Dry and wet	0.9–4.5% [a] 3.7% [b]
Total porosity			Wet 17–22% [a] Dry 18–20% [a] Dry and wet – 21.1% [b] Wet – 20.5%

[a] Kusterle 1992.
[b] Cornejo-Malm 1995.
[c] Blasen 1998 (average values from 337 samples).

hardening and the curved shape of the stress–strain graph for concrete in compression (see Figure 2.5) – i.e. the tangent modulus decreases with increasing strain. Clearly, the higher the initial level of porosity in the concrete, the higher the initial local stresses will be. Above an applied stress of about 70% of the maximum compressive strength, cracking occurs within the paste, and the microcracks start to join up (Rokahr & Lux 1987). After the maximum compressive strength has been reached, macrocracks form

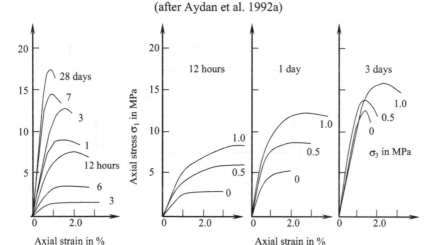

Figure 2.5 Stress–strain curves for sprayed concrete at different ages (after Aydan et al. 1992a).

as the microcracks localise in narrow bands, and the load that the concrete can sustain decreases (Feenstra & de Borst 1993).

Under triaxial compressive stresses, the cracking may be suppressed by the lateral stresses, and, if the confinement is high, the mode of failure is crushing. Hence, the maximum compressive stress under triaxial compressive loading is much higher than the uniaxial or biaxial strength (Chen 1982, Neville 1995). However, in the case of a tunnel lining, the stress state is largely biaxial since the radial stresses in the lining are much lower than the tangential and longitudinal stresses (Meschke 1996). In compression, the biaxial strength is only 16% greater than the uniaxial strength, when $\sigma_2/\sigma_1 = 1.0$, and 25% greater, when $\sigma_2/\sigma_1 = 0.5$ (Chen 1982) – see Figure 2.6. In intermediate states of stress between pure compression and pure tension, the presence of a tensile stress reduces the maximum compressive stress attainable (Chen 1982). In the biaxial stress case, it is often assumed that the maximum compressive stress reduces linearly from the uniaxial value (when the tensile stress is zero) to zero (when the tensile stress equals the maximum uniaxial tensile stress) – see Figure 2.6.

To summarise, the strength of concrete depends on one hand on the strength of the main components – the hardened cement paste and aggregate – and on the other hand on the density of the sample. Strength rises with age since the quantity of hardened cement paste increases with age as the hydration process continues and the quantity of voids decreases, rather than because of any actual change in the mechanical properties of the microscopic constituents (Ulm & Coussy 1995). This increase in

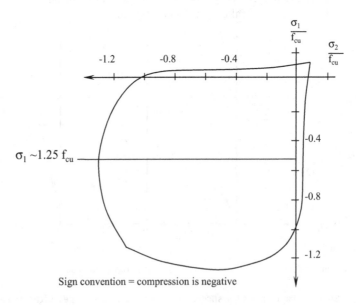

Figure 2.6 Biaxial strength envelope (Chen 1982).

strength with time can lead to strengths well above the specified grade (e.g. see Table 2.4) which is both unnecessary and may even be a disadvantage. This can lead to the phenomenon of embrittlement in steel fibre reinforced concrete (see Section 2.2.2). On a microscopic level, the local stress depends on the effective area of solid material sustaining the stress (if one ignores any contribution of pore water pressure), and the growth of cracks then depends on the strength of the bond between the hardened cement paste and aggregate compared to these local stresses. From this simplified theory of how concrete behaves under compressive loading, one could conclude that, to improve the strength of a concrete, one should improve the density of the material, by maximising hydration and minimising the porosity, and one should also improve the hardened cement paste–aggregate interaction.

Influences on behaviour

Modern specifications typically require compressive strengths at 28 days of 20 MPa or more for temporary sprayed concrete (Brooks 1999) and 30 MPa or greater (e.g. C35/45 acc. to ÖBV (2013)) for permanent concrete. Project specific requirements may lead to even higher strengths. The sprayed concrete must also possess sufficient adhesion to adhere to the ground and to support load, from the ground as well as other sources, such as blasting, soon after it has been sprayed. Hence, in contrast with conventional concrete, the sprayed concrete mix must be designed to attain a relatively high early compressive strength (see Figure 2.2) as well as meeting the long-term criteria. Furthermore, the mix must meet more stringent workability and pumpability criteria than conventional concrete. Of these competing criteria, traditionally the early age strength (which determines the thickness of layers that can be formed and the safety of the tunnel heading) and the pumpability requirements have dominated, at the expense of optimising the longer term strength (Kusterle 1992, Darby & Leggett 1997).

Accelerators – accelerating the hydration reaction will increase the strength of the sprayed concrete at early ages (see Figure 2.2) – see also 2.1.1 and 2.1.5. Traditionally, high early strengths have been achieved by adding accelerators to the mix in the spraying nozzle. This has several disadvantages. Firstly, accelerating the hydration reaction causes more, smaller hydrated calcium-silicate crystals to grow. A slower reaction permits larger crystals to grow, resulting in higher strengths in the long term (Fischnaller 1992, Atzwanger 1999).[2] Secondly, many of the early accelerators were very alkaline and hazardous to the health of workers in the tunnel. Some accelerators, such as water-glass (sodium silicate), not only led to low strengths at 28 days but due to their instability, the strength actually decreased with age (Kusterle 1992). The concerns over low long-term strengths and health and safety have forced the introduction of new accelerators – so-called "alkali-free" or "low-alkali" accelerators (Brooks 1999). With these new products and other new additives, the compressive strength gain of sprayed concrete can be controlled with a fair degree of accuracy and tailored to suit the particular requirements of the project. Together with additives such as microsilica, the competing demands of high early strength and long-term strength can be met more satisfactorily (see Table 2.4).

Cement and cement replacements – ordinary Portland cement is normally used. "Spray cements" and calcium sulphoaluminate cements can be used in the dry mix process – see Section 2.1.1.

Water–cement ratio – the lower the water–cement ratio, the higher the strength because fewer voids are left after hydration. Complete hydration of cement requires a water–cement ratio of approximately 0.23. However, pumpability requirements dictate that higher water–cement ratios are used for wet mix sprayed concrete than for cast concrete. In the dry mix process, the water–cement ratio is controlled by the nozzleman. Typically, average values are 0.3 to 0.55 for dry mix; the ratio for wet mixes lies in the range of 0.4 to 0.65 (ITA 1993).

Grading curve and aggregate – the maximum aggregate diameter is usually limited to about 10 to 12 mm, compared to around 20 mm for cast concrete. Strength increases with increasing maximum diameter of aggregate but, the larger the pieces of aggregate, the more of them are lost in rebound (Kusterle 1992, Brite Euram 1998, Austin et al. 1998). As a whole, the grading curve for sprayed concrete is biased towards the finer end (see Figure 2.1) for ease of pumping (Norris 1999). A well-graded aggregate curve is essential for good pumpability. Both crushed and round gravel can be used as aggregate. Some experimental evidence suggests that the type of gravel causes little difference in the quality of the sprayed concrete (Springenschmid et al. 1998). However, anecdotal evidence from various sites suggests that the grading curve and sometimes the type of aggregate too may have a great influence on the sprayed concrete. Aggregate that has a smooth grading curve should be used and angular particles should be avoided since they are more difficult to pump. If necessary, either the sand or the aggregate may be angular, but one of the two should be rounded.

Microsilica – the addition of microsilica has two main advantages. Firstly, it improves the adhesion of the sprayed concrete, permitting accelerator dosages to be reduced or thicker layers of sprayed concrete to be placed. The higher adhesion reduces dust and rebound (Brite Euram 1998). Secondly, acting as a very reactive pozzolanic pore-filler, microsilica improves the long-term density, which is beneficial for strength and durability. In general, microsilica enhances the quality of the sprayed concrete, improving durability as well as mechanical properties (Kusterle 1992, Norris 1999). The main disadvantage is its high water demand, which requires more plasticiser or water or both (Norris 1999, Brooks 1999). In fact, it has been suggested that, in the case of dry mix sprayed concrete, this additional water may be partially responsible for the reduction in dust and rebound (Austin et al. 1998) as well as pumpability. NCA (2011) recommends adding microsilica at a dosage of between 4 and 10% bwc.

Fibres – contradictory evidence exists as to whether the addition of steel fibres alters the compressive strength of sprayed concrete. Vandewalle (1996) suggests that they have little beneficial effect, while Brite Euram (1998) suggests that steel fibres increase the compressive strength by 10 to 35%. Polypropylene fibres were also found to enhance strength but they also increase the water demand so that there is little overall benefit (Brite Euram 1998). Generally, it is assumed that fibres have no beneficial effect on the compressive strength.

Other additives and admixtures – individually plasticisers, stabilisers and other additives may not have a detrimental impact on the mechanical properties of sprayed concrete, but one must always be aware that combinations of accelerators and additives may produce unfavourable results, such as significantly reduced strengths (Brite Euram 1998). Compatibility testing before work begins on site is used to identify such unfavourable combinations.

Anisotropy – concrete is not naturally anisotropic, and the anisotropy seen in sprayed concrete is a consequence of the way in which it is produced. Compressive strengths have been found to be 10–25% higher in the plane perpendicular to the direction of spraying (Cornejo-Malm 1995, Huber 1991, Fischnaller 1992, Bhewa et al. 2018).[3] However, others have reported no variation in strength with direction of testing (Purrer 1990, Brite Euram 1998). At first sight, higher strengths perpendicular to the direction of spraying may seem paradoxical since the spraying jet is the sole means of compaction for sprayed concrete. This "softer response" may be due to compaction at the less dense interfaces between layers of sprayed concrete (Aldrian 1991). The strength is normally tested in the direction of spraying, since the samples are usually cored from sprayed test panels or the lining itself, whereas the major compressive stresses are in the plane perpendicular to this (Golser & Kienberger 1997, Probst 1999). Hence the use of strength values from cores could be considered as conservative. Steel fibre reinforced sprayed concrete exhibits pronounced anisotropy in its behaviour under both compression and tension (see Section 2.2.2). Normally the anisotropy of the sprayed concrete (and indeed the stiffening effect of the layers of mesh or fibres) is ignored.

Temperature – Cervera et al. (1999a) proposed a reduction factor for the ultimate compressive peak stress, to account for the effects of (constant) elevated ambient temperatures during curing. The reduction factor is $k^{iso} = [(100-T^{iso})/(100-20)]^{nT}$, where $nT = 0.25$ to 0.4 and T^{iso} is the (constant) temperature during hydration. This gives comparable reductions to those found experimentally by Seith (1995), e.g. 25% reduction in strength for curing at 60°C compared to curing at 16°C.

In conclusion, if properly produced, sprayed concrete can achieve high early strengths and long-term strengths (see Table 2.4). The exact shape of the strength gain curve will depend on the sprayed concrete mix and additives. Because of the interest in early age strength gain, several authors have

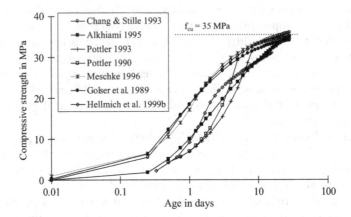

Figure 2.7 Predictions of strength development.

proposed equations that can be used to relate the compressive strength to age (e.g. Aldrian 1991, Chang 1994, Alkhiami 1995, Yin 1996 (after Weber 1979), Pöttler 1990, Meschke 1996) – see Figure 2.7 and Appendix A. While these predictions may match data well in general, they tend to under-estimate the strengths at very early ages (i.e. < six hours). Other more complex approaches have been developed to include ageing in numerical analyses (see Section 5).

2.2.2 Strength in tension

This section covers the tensile strength of both plain and reinforced sprayed concrete.

Theories and mechanisms

Even more so than in the case of compression, when under tension, cracking governs the behaviour. Up to 60% or more of the maximum uniaxial tensile stress, few new microcracks are created, and so the behaviour is linearly elastic (Chen 1982). The period of stable crack propagation under tension is shorter than compression. At about 75% of the maximum uniaxial tensile stress, unstable crack propagation begins, and a few cracks grow rapidly until failure occurs. The exact cause of tensile rupture is unknown but it is believed to originate in flaws in the hardened cement paste itself and at the paste/aggregate interface, rather than in the voids and pores, although these features contribute to the formation of stress concentrations (Neville 1995).

Normally, the tensile strength of concrete is ignored in design because it is low – typically about one-tenth of the compressive strength – and because of the brittle nature of the failure once the maximum is reached. To counteract this, tensile reinforcement is added to concrete. Reinforcement in sprayed concrete tunnel linings is normally by steel mesh, steel fibres or macrosynthetic fibres, although experiments have been performed with other materials as fibres, such as ("soft") polypropylene fibres (Brite Euram 1998), basalt (Sandbakk et al. 2018), glass fibre reinforced polymer (Ansell et al. 2014, Sandbakk et al. 2018) and even hemp (Morgan et al. 2017).

When bar reinforced concrete is loaded with a tensile stress, cracking occurs in the concrete as before. However, the bond between the uncracked concrete and the steel bars permits a gradual transfer of the tensile load from the cracking concrete to the steel as the load increases (see Figure 2.8). The reinforced concrete continues to act as a composite, and hence it has a stiffer response to loading than the reinforcement or concrete alone. This phenomenon is known as "tension stiffening" (Feenstra & de Borst 1993).

Fibre reinforcement

Fibre reinforcement has a similar effect, although the interaction between the fibres and the matrix is more complex. Fibre reinforcement of concrete is a big topic in its own right, and more detailed information can be found elsewhere (e.g. EN 14889-1, EN 14889-2, Thomas 2014, ITAtech 2016) – see Table 2.6 for typical properties of the fibres. In short, the fibres bridge the opening cracks, thereby continuing to carry tensile forces across the cracks. The fibres are usually deformed in some manner to improve their resistance to being pulled out of the concrete as a crack opens. High grade steel is used for the fibres (typically yield strengths around 1,000 MPa) so that failure occurs by means of a "ductile" process in which individual

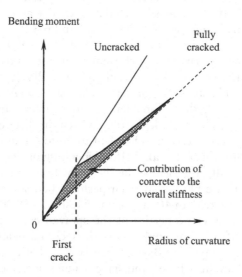

Figure 2.8 Tension stiffening of reinforced concrete.

Table 2.6 Typical properties of structural fibres for sprayed concrete

	Steel	Structural synthetic
Density (kg/m³)	7800	900
Fibre length	30–50	40–60
Aspect ratio	30–65	50–90
No. of fibres per kg	14,500	35,000
Elastic modulus (kN/mm²)	200	5.0–9.5
Tensile strength (N/mm²)	800–1,500	500–550

fibres are pulled out of the concrete. This can impart significant tensile capacity to the concrete, even at early ages (i.e. two days) (Bjontegaard et al. 2014). If the concrete is too strong and it prevents pull-out, the individual fibres would snap and the overall failure process would be a brittle one (Bjontegaard et al. 2014, Bjontegaard et al. 2018). This process has been termed "embrittlement" – Bernard (2009, 2014). Since concrete continues to hydrate beyond the standard age for testing (28 days), this phenomenon can affect mixes which might appear fine on paper. This can be avoided by using higher strength steel or macrosynthetic fibres (Bjontegaard et al. 2014, ITAtech 2016). The typical dosage for steel fibres ranges from 20 to 60 kg/m³. In practical terms fibres tend only to be used with wet mix sprayed concrete. The difficulties in mixing fibres in dry mix lead to excessive rebound.

The first non-metallic fibres to be widely used in sprayed concrete were soft polypropylene fibres as noted above. These fibres did not have a meaningful beneficial effect on tensile behaviour except at an early age when they can resist shrinkage (see Sections 2.2.6 and 6.8.9). A new generation of structural plastic fibres has emerged during the 1990s, with mechanical properties that can rival the performance of steel fibre reinforced shotcrete (SFRS) (Tatnall & Brooks 2001, Hauck et al. 2004, DiNoia & Rieder 2004, Denney & Hagan 2004, Bernard et al. 2014, ITA 2016). Pioneered in the mining industry these fibres have similar dimensions to steel fibres, but they are lighter, softer and less strong. Given the difference in density, a dosage of 5 to 10 kg/m³ can be equivalent to 30 to 40 kg/m³ of steel fibres (Melbye 2005, Bjontegaard et al. 2018), depending on the exact properties of the fibres. Typically, the yield strengths of the fibres are greater than 500 MPa. Macrosynthetic fibres may even perform better than steel fibres at large deflections, more than 40 mm in panel tests (BASF 2012, Bjontegaard et al. 2014). The other properties of the sprayed concrete, such as pore structure and stiffness, are similar in both steel and macrosynthetic fibre reinforced samples (Myren & Bjontegaard 2014). They also cause less wear on equipment than steel fibres and have a much lower carbon footprint (Bernard 2004a) – see also Figure 2.4. There tends to be less rebound when using macrosynthetic fibres (Kaufmann & Frech 2011). Crehan et al. (2018) describe the measures needed to prevent any environmental contamination by macrosynthetic fibres. Filters may also need to be fitted to water pumps to prevent the fibres from blocking them.

Experiments have also been conducted recently with textile glass fibre meshes made from alkali resistant glass. At present these meshes can only be used in gunite with a maximum size of aggregate of less than 2 mm due to the close spacing of the threads in the mesh (Schorn 2004). Glass fibre reinforced polymer (GFRP) meshes with similar dimensions to steel meshes have been using in some mining applications.

Influences on behaviour

The tensile strength of sprayed concrete is subject to the same influences and can be improved in the same ways as the compressive strength (see Section 2.2.1). A spin-off of this is that the tensile strength of concrete can be reliably estimated from the compressive strength. Various empirical formulae have been proposed (see Appendix C). It is generally assumed that the tensile strength increases with age at the same rate as compressive strength.

Reinforcement – bars and fibres

Sprayed concrete tunnel linings are often formed by spraying several layers of concrete. Mesh or bar reinforcement is placed on the surface of the last layer sprayed and encased in concrete by the next layer (see Figure 2.9). Shadowing may adversely affect the bond between the steel and the concrete as well as its durability (see Section 2.2.9).

SFRS has many advantages (Brite Euram D1 1997, Vanderwalle et al. 1998):

- SFRS can behave in an almost elastic perfectly plastic manner (Norris & Powell 1999), withstanding very large post-yield strains.
- They can be included in the sprayed mix, reducing the cycle time and improving safety, since there is no mesh to be fixed at the face.
- Fibres are more effective in controlling shrinkage cracking than typical mesh or bar reinforcement.
- Corrosion of the fibres is not generally thought to be a significant problem (Nordstrom 2001, ACI 544.5R-10 2010, Nordstrom 2016, Hagelia 2018), except in very thin linings, and there are no problems of shadowing.

These qualities make SFRS popular for support in tunnels in blocky ground. They are also desirable in tunnels in high stress environments, where large deformations are expected and SFRS can be used in conjunction with rock bolts or in tunnels with permanent sprayed concrete linings,

Figure 2.9 "Shadowing" in sprayed concrete behind reinforcement.

which are acting mainly in compression (Annett et al. 1997, Rose 1999, Thomas 2014).

Although the fibres orientate themselves mainly in the plane perpendicular to the direction of spraying (e.g. Cornejo-Malm 1995, Norris & Powell 1999), the moment capacity of SFRS is quite small, at typical fibre dosages (Thomas 2014). If a large moment capacity is required, bars or mesh reinforcement are needed. This can be the case in soft ground tunnels, especially at junctions. By optimising the shape of the tunnels, bending moments can be kept small, permitting the use of fibre reinforcement (e.g. Heathrow Terminal 5 and Crossrail projects in the UK). Various design guidance has been proposed for fibres (DBV 1992, RILEM 2003 and most recently fib 2010) – see Thomas (2014) for a review of them. As an aside, fibres are widely used in another branch of tunnelling – segmental linings (ITAtech 2016).

Because the fibres orientate themselves in the plane perpendicular to the direction of spraying, when tested in compression in this plane, SFRS exhibits a stiffer response pre-peak, higher peak stresses and a softer post-peak response, compared to tests performed in the same direction as spraying (Brite Euram 1998). The axial and lateral strains at ultimate stress are lower for the same reason.

The use of non-metallic fibres removes the residual concern over durability (see also section 2.2.9) but does introduce a new question – namely, creep. Sprayed concrete with macrosynthetic fibres can behave in a similar way to steel fibres when the load is less than 40% of the peak load (Bernard 2010 and see Figure 2.10), and creep does not seem to reduce the ultimate energy absorption (Bernard 2004a). Another detailed study found no substantial difference between sprayed concrete reinforced steel fibres,

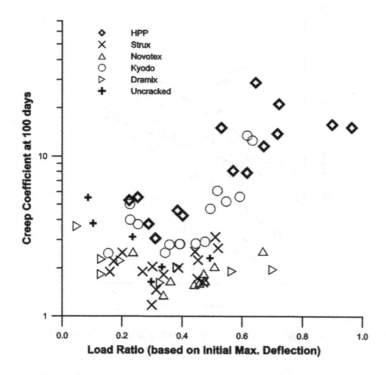

Figure 2.10 Creep coefficient at 100 days expressed as function of load ratio (based on residual load capacity at initial maximum deflection) (Bernard 2010).*

macrosynthetic fibres and welded wire mesh during creep loading over one year when loaded up to 50% of the peak capacity (Larive et al. 2016). However, some studies suggest it has a higher creep coefficient, possibly as high as twice as that of steel fibre reinforced sprayed concrete (Bernard 2004a, MacKay & Trottier 2004), albeit the creep coefficient of the samples depends heavily on the type of macrosynthetic fibre, the dosage and load intensity. In practice this does not appear to present any problems (Plizzari & Serna 2018), but designers should consider each case on its own merits – see also Section 2.2.7.

2.2.3 Strength in other modes of loading

The input parameters for concrete models, such as Drucker–Prager or Mohr–Coulomb plasticity models, are generally derived from the compressive and tensile strengths. Unlike soils, the **shear strength** of sprayed concrete is not normally tested directly. However, the shear strength may be critical to the performance of the sprayed concrete lining (Barrett & McCreath 1995, Kusterle 1992), particularly if the lining thickness is very small (NCA 1993). Information on the shear strength of sprayed concrete

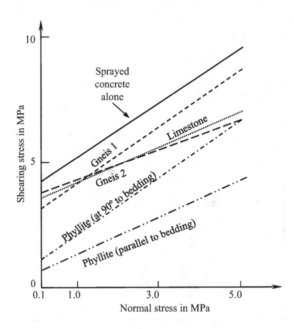

Figure 2.11 Bond strength in shear to various substrates (after Kusterle 1992).

bonded to various rocks is contained in Figure 2.11 (see also NCA (1993))
for information on bond strengths to rock).

Similarly, the **bond strength** of sprayed concrete (both to the substrate
and between successive layers of sprayed concrete) is important to the performance of sprayed concrete (Sjolander et al. 2018). In rock tunnels, when
considering the lining, acted upon by a single wedge, the failure of sprayed
concrete linings has been found to occur in two stages – usually by debonding, followed by failure in flexure (Barrett & McCreath 1995, Sjolander
et al. 2018). Table 2.7 contains indicative values for the peak strengths of
sprayed concrete in other loading modes. The bond strength between layers
of sprayed concrete has been examined in the context of permanent sprayed
concrete linings, in which the final layer may be added months after the
first (Kusterle 1992, Brite Euram 1998). Typical values of bond strength
between layers of sprayed concrete range between 0.8 and 2.6 MPa (Brite
Euram 1998). Bryne et al. (2011) reported bond strengths to rock ranging
from 0.1 to 3.0, depending on the strength of the rock, with an average in
normal condition around 1.0. Clements et al. (2004) reported values for the
bond of sprayed concrete to rock ranging from 2.83 to 11.3 MPa at 150
days. Jolin et al. (2004) report much larger values of more than 20 MPa for
bond strengths to reinforcing bars at an age of 28 days. Provided that the
substrate has been cleaned well, acceptable bond strengths can be achieved.
Having said that, the failure is generally at the contact with the rock or in

the rock, rather than in the concrete itself (Clements et al. 2004). Ansell (2004) reported results from Malmgren and Svensson that showed how the adhesion increases with age from starting values of between 0.125 to 0.35 MPa (when sprayed) to between 1.0 and 1.4 MPa (at 28 days), but, in more recent research by the same team, they reported bond strengths of about 0.2 MPa at ten hours, rising more rapidly to an average value of 1.5 MPa at three days (Bryne et al. 2014). Sjolander et al. (2018) cite test evidence that the bond strength is independent of the thickness of the lining (at least for thicknesses from 30 to 140 mm), but the face plates for rock bolts can restrict debonding. Lamis and Ansell (2014) proposed the following equation for the increase in bond strength with age:

$$f_{cb} = 2.345e^{-0.858t^{-0.97}}$$

(2.2)

Where f_{cb} is the bond strength in MPa and t is the age in days.

In soft ground, the tunnel lining is subjected to a more even loading than in rock tunnels so the bond strength may not be an important parameter for soft ground tunnels.

2.2.4 Stress–strain relationship in compression

Behaviour and influences

The mechanisms behind the stress–strain behaviour of concrete in compression have been described already in Section 2.2.1. A stress–strain curve for a uniaxial test typically shows a linear elastic response up to the limit of proportionality, followed by what becomes an increasingly softer response as the maximum compressive strength is approached (see Figure 2.5). After reaching a peak value, the stress that can be sustained falls with increasing strain until the ultimate compressive strain is reached and the sample fails completely. In fact, the onset of failure may occur before the peak stress, since the maximum volumetric strain is reached at a stress of between 0.85 and 0.95 of the peak, and after this point dilation starts (Brite Euram C2 1997). The observed shape of the post peak descending branch of the stress–strain curve depends heavily on the confinement and the boundary conditions imposed by the experimental equipment, due to the localisation of cracking (Swoboda et al. 1993, Choi et al. 1996). For that reason, one could describe concrete as being a "near-brittle" material and ignore the post peak region, concentrating rather on the pre-peak region. Generalised mathematical relationships have been developed for this region (e.g. Eurocode 2 2004, BS8110 Part 2 1985) that agree well with a large range of uniaxial, biaxial and triaxial data, including tests on sprayed concrete (Brite Euram C2 1997).

The stress–strain behaviour of concrete under multi-axial stress states is very complex. While the increase in compressive strength has been clearly

Table 2.7 Strength in other modes of loading (after Barrett & McCreath 1995)[a]

Strength	8 hours	I day	7 days	28 days
"Poor" bond strength in MPa	–	–	–	0.5
"Good" bond strength in MPa	–	–	1.5	2.0
Direct shear strength in MPa	1.0	2.0	6.0	8.0
Flexural strength, $f_{fck,fl}$, in MPa	1.0	1.6	3.4	4.1
Diagonal tensile strength in MPa	0.75	1.0	1.75	2.0
Uniaxial compressive strength in MPa	5.0	10.0	30.0	40.0

[a] All sprayed concrete strengths are for an unreinforced mix with silica fume added.

established, it is more difficult to form a definitive picture of the strain behaviour since it depends heavily on the boundary conditions in the experiments (Chen 1982). That said, increasing the confining pressure appears to lead to more ductile behaviour (Michelis 1987, Aydan et al. 1992a) – see Figure 2.5. Triaxial behaviour will not be discussed further since this stress state in a tunnel lining is basically biaxial. The effect of tension in mixed biaxial loading is to reduce the peak (and failure) principal compressive and tensile strains (Chen 1982). The maximum strength envelope under biaxial loading can be considered to be independent of the stress path (Chen 1982).

In most cases the stress level in a tunnel lining is relatively low. Considering a typical tunnel in soft ground, where the principal stresses in the lining might be 5.0, 5.0 and 0.5 MPa and the 28-day strength is 25 MPa, the normalised octahedral mean stress (σ_{oct}/f_{cyl}) is only 0.14. Hence one can ignore those effects, which occur at moderate to high stress levels, such as the curved nature of the yield surface meridians (see Figure 2.12).

Particular points of interest to the designer are the initial elastic modulus, the limit of proportionality (i.e. limit of elastic range), the peak stress and strain. The behaviour post peak and at high stress–strength ratios (e.g. > 0.85) will not be discussed further here on the grounds that structures are not normally designed to operate in this region.

Elastic region

Elastic limit – the behaviour of sprayed concrete at an early age, in compression tests, has been characterised as viscous (from zero to one or two hours old), visco-elastic (1 to 11 hours) and elastoplastic (from 11 hours onwards) (Brite Euram 1998). This behaviour may vary depending on the level of loading. Figure 2.13 shows how the ratio of yield stress to peak stress for sprayed concrete (estimated visually from stress–strain curves or from published data from (Aydan et al. 1992a)) varies with age. Some data suggest that the yield point is relatively high – 0.70 to 0.85 of the peak stress (Aydan et al. 1992a).

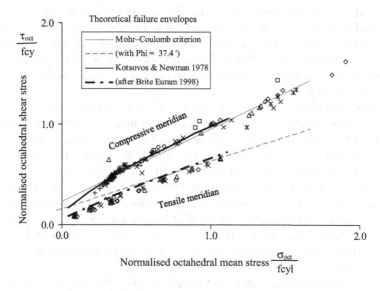

Figure 2.12 Normalised octahedral stress envelope for sprayed concrete (with published data from Aydan et al. 1992a, Brite Euram 1998, Probst 1999).

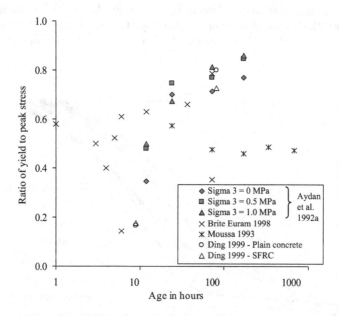

Figure 2.13 Yield stress–peak stress ratio (published data, including data from triaxial tests by Aydan et al. 1992a at various confining pressures – sigma 3).

Other data suggest much lower ratios, tending towards the generally accepted yield ratio for mature concrete of 0.3 to 0.4 (Chen 1982, Feenstra & de Borst 1993). If one examines tests that included unloading/reloading cycles (e.g. Moussa 1993, Probst 1999), one can see that all the strain is not recovered upon unloading even at low stresses. This supports the view that the elastic limit is low.

Elastic modulus – considerable data exists for the elastic modulus (calculated from uniaxial compression tests) and how it varies with age (e.g. Chang 1994, Kuwajima 1999). The modulus grows rapidly with age in a similar way to compressive strength, although it appears to grow at a faster rate (Byfors 1980, Chang 1994). In one study, it was found that the stiffness of sprayed concrete samples was significantly lower than the values predicted by equations from normal concrete standards at ages of one day and more (Galobardes et al. 2014). This was attributed to the effects of spraying, and a correction factor was proposed which would reduce the predicted values by about 20% for typical values of porosity and rebound.[4]

Various formulae have been proposed to relate the elastic modulus to age (see Figure 2.14 and Appendix A). Other more complex approaches have been developed to include ageing in numerical analyses (see Section 5). Sprayed concrete may exhibit anisotropy, with the elastic modulus in the plane perpendicular to the direction of spraying being higher than in the plane parallel to the direction of spraying. Celestino et al. (1999) report that it is 40% higher while Cornejo-Malm (1995) and Bhewa et al. (2018)[5]

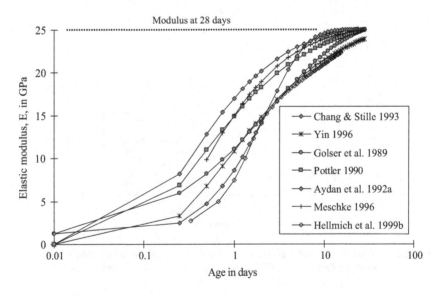

Figure 2.14 Predictions of the development of elastic modulus with age.

recorded an increase of about 10% (see also Section 2.2.1). Such anisotropy is generally ignored in design.

Ansell (2004) reported that Nagy proposed that the dynamic elastic modulus, E_{dyn}, is related to the static modulus by the following equation:

$$E_{dyn} = E.\left(1 + \eta^{0.35}\right) \tag{2.3}$$

where $\eta = 0.14$ for very young concrete and decreases to 0.05 at an age of two days.

Ultrasonic measurements have been used as a non-destructive method of measuring the elastic modulus (e.g. Gibson & Bernard 2011, Galan et al. 2018). Bhewa et al. (2018) reported that the dynamic modulus can be determined from ultrasonic measurements by the following equation:

$$E_{dyn} = \frac{(1 + v) \cdot (1 - 2v)}{(1 - v)} \cdot \gamma \cdot V^2 \tag{2.4}$$

where v is Poisson's ratio, γ is the concrete density and V is ultrasonic longitudinal wave velocity. In their tests, they found little anisotropy in the dynamic elastic modulus (i.e. < 5%), in contrast to the static modulus, but as before, the modulus was consistently higher perpendicular to the direction of spraying.

Bhewa et al. (2018) considered the relationship between the static and dynamic moduli proposing a simple equation for this. Examining their raw data, the following equation appears to match better:

$$E_{static} = 0.750 * E_{dyn} \tag{2.5}$$

Since concrete behaves linearly elastic up to a limit of about 0.4 times the uniaxial compressive strength, the elastic modulus should be determined from loading within this range only.

Poisson's ratio – within the elastic range and up to 80% of the maximum stress, Poisson's ratio remains constant for mature concrete, ranging between 0.15 and 0.22 and with an average of about 0.2 (Chen 1982). The actual value of the Poisson's ratio depends mainly on the type of aggregate, with lower values in concrete with lightweight aggregate (Neville 1995). Mature sprayed concrete exhibits the same behaviour, but there is some evidence that the Poisson's ratio varies with age. Kuwajima (1999) measured the dynamic Poisson's ratio using ultrasound and found that it decreased with age from close to 0.5 to about 0.28 (see Figure 2.15). Bhewa et al. (2018) reported similar results for samples cored parallel to the direction of spraying but a smaller variation for samples cored perpendicularly. Dynamic Poisson's ratio values are usually higher than static ones (Neville 1995). Aydan et al. (1992a) and Aydan et al. (1992b) report a similar variation with age. They measured values of Poisson's ratio close to 0.45 initially

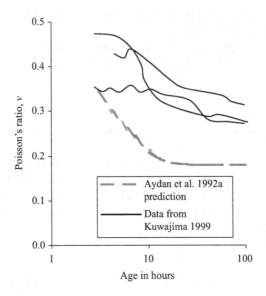

Figure 2.15 Variation of Poisson's ratio with age.

falling to about 0.2 at 12 hours, but they do not state how these values were obtained (see Appendix A for the equation relating Poisson's ratio to age). Plane strain compression tests from the Brite Euram project (Brite Euram 1998) suggest that the Poisson's ratio at early ages (3 to 16 hours) is closer to the mature values.

Plastic region up to peak stress

Sprayed concrete can withstand very large plastic strains at an early age. The strain at peak stress decreases with increasing age (see Figure 2.16), from as high as 5.0% at one hour old to a relatively constant value of 1.0%, from 100 hours onwards. The peak strain of mature concrete is normally assumed to be about 0.3% in uniaxial and biaxial loading (Chen 1982, BS8110 Part 1 (1997)). The ultimate strain (at failure) also decreases with age (see Figure 2.17), and the behaviour becomes more brittle (see Figure 2.5) (Swoboda et al. 1993). Swoboda and Moussa (1994) observed a similar trend when plotting a graph of maximum strain against the logarithm of the compressive strength of the sprayed concrete rather than its age. It is believed that the deformation behaviour of mature sprayed concrete is not affected much by changes in mix constituents (Brite Euram 1998). For example, the normalised stress at maximum volumetric strain does not change with variation in accelerator

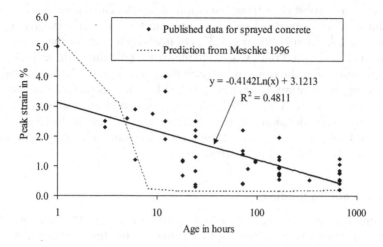

Figure 2.16 Peak compressive strain vs. age.

dosage, and the strain at peak stress is also independent of accelerator dosage (Swoboda & Moussa 1992).

The addition of steel fibres appears to make the sprayed concrete more ductile in compression, with a strain at peak stress of about 0.42% (28 days), compared to 0.20% for plain sprayed concrete (Brite Euram 1998).

Unloading

In uniaxial compression tests, the unloading (and reloading) modulus is stiffer than the initial loading modulus (Michelis 1987, Probst 1999, see

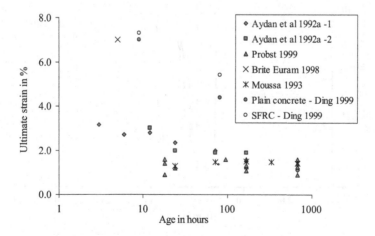

Figure 2.17 Ultimate compressive strain vs. age.

Figure 2.18). Probst (1999) suggests multiplying the current value of (initial) modulus by a factor of 1.1 to 1.5, to account for this, while the average of results from Aldrian (1991) was 1.27. For loads up to 70% of the peak stress, Moussa (1993) found that, when reloaded after unloading, the stress–strain path would rejoin the original curve at the point where it had departed on unloading and continue as if the unloading had not occurred (as one would expect for concrete (Chen 1982)). Using a new type of testing rig, which does not require demoulding and therefore may reduce disturbance of the samples, Probst (1999) observed the same behaviour as Moussa in uniaxial tests, even up to 80% of peak stress.

Damage due to loading

The question of whether or not early loading damages the concrete is of great importance to permanent sprayed concrete linings (see Section 4.2.5). Little research has been done on this aspect of sprayed concrete. Moussa (1993) concluded from experimental work that stresses below a utilisation factor, α, of 70% of the current peak stress had no detrimental effect on the later peak stress. He proposed a linear relationship between the reduction factor and the stress above this level:

$$R_{\mathrm{dfc}} = 2.532\,(\sigma \,/\, f_{ct1} - 0.69) \qquad\qquad (2.6)$$

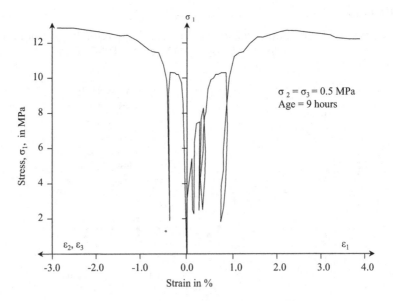

Figure 2.18 Compressive test on sprayed concrete.

which ranges from 0 at $\alpha = 0.69$ to 0.78 at $\alpha = 1.0$, where $\alpha = \sigma/f_{c\,t1}$ and $\sigma =$ the stress applied and f_{ct1} = the uniaxial strength at the age of loading, t1. However, the experimental data were quite scattered (Swoboda et al. 1993). By comparing the strength of samples from creep tests with samples from parallel shrinkage tests, Huber (1991) found that the strength of loaded samples was 80% of the unloaded ones. The utilisation factors in the creep tests ranged from 20 to 70%. Chen (1982) suggested that for normal concrete unstable crack propagation occurs when utilisations exceed 75%.

2.2.5 Stress–strain relationship in tension

The mechanisms behind the stress–strain behaviour of unreinforced concrete in tension have been described already in Section 2.2.2. Experimental data on conventional concrete in tension are scarce, especially for concrete at early ages. A stress–strain curve for a uniaxial test on mature concrete typically shows a linear elastic response up to 60% of the maximum stress (Chen 1982). As more and more microcracking occurs, the response becomes softer until the maximum stress is reached. After the peak stress, the stress quickly drops to zero for unreinforced concrete. The precise nature of the descending branch of the stress–strain curve depends heavily on the arrangement of the testing rig (Hannant et al. 1999, Chen 1982). Reinforcement enables tensile forces to be carried even though the concrete has cracked, as discussed earlier (see Section 2.2.2 and Figure 2.8). The utilisation factor in parts of a tunnel lining under tensile stress is likely to be much higher than in areas of compressive stress because the tensile strength is much lower.

At very early ages (i.e. less than four hours old), cast concrete appears to behave plastically and can be strained by up to 0.5% or more (Hannant et al. 1999). However, this ultimate strain reduces sharply with increasing age and is about 0.05% at five hours.

Sprayed concrete exhibits the same behaviour (see Figure 2.19). In uniaxial tensile tests, plain sprayed concrete and fibre reinforced sprayed concrete (SFRS) behave similarly (Brite Euram 1998) and, as for plain concrete, the ultimate strain reduces sharply in the first few hours. The effect of the fibres can be seen in Figure 2.19 as converting an otherwise brittle failure into a more ductile one, in which the stress–strain curve descends slowly from the peak – see also Section 2.2.2. A similar effect is observed in flexural tests on SFRS beams. Deformation hardening can even be achieved with the right mix and type of high strength steel fibre (ITAtech 2016). In the context of soil–structure interaction and the structurally redundant shell structures of tunnel linings, it is not clear that deformation hardening is required, and almost all the fibre reinforced linings that have been constructed to date and continue to perform well have used deformation-softening fibre reinforced concrete (Thomas 2014).

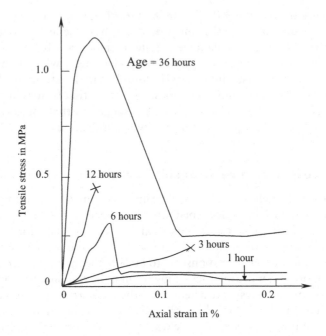

Figure 2.19 Uniaxial tensile tests on samples of mix IK013 at different ages (Brite Euram 1998).

The elastic modulus under tension is also assumed to be equal to that under compression for unreinforced concrete. Poisson's ratio is the same in tension as in compression in the elastic region.

2.2.6 Shrinkage and temperature effects

The following sections will cover the various forms of shrinkage (plastic, autogeneous, drying and carbonation shrinkage) and temperature effects that induce strains in sprayed concrete linings.

Shrinkage

Plastic shrinkage is the contraction caused by the loss of water from the fresh concrete's surface due to evaporation or suction, by adjacent dry soil or existing concrete, while the concrete is still plastic (Neville 1995). If the water lost exceeds the volume brought to the surface by bleeding, surface cracks may appear. Plastic shrinkage increases with increasing evaporation, cement content and water–cement ratio and decreases with a decreasing tendency for bleeding. A typical value for (linear) shrinkage after 24 hours is 0.2% (for 400 kg/m³ cement, air temp = 20°C, relative humidity = 50%, air velocity of 1.0 m/s – Neville 1995).

Plastic settlement also occurs in the first hours after casting and is sometimes confused with plastic shrinkage. Plastic settlement is caused by differential settlement of the concrete over obstructions such as large aggregate or reinforcement (Neville 1995). Plastic shrinkage is the early part of drying shrinkage (see later) that occurs while the concrete is still plastic. Two key factors influencing this are the degree of compaction and the rate of build-up of concrete. In the case of sprayed concrete, in the crown of the tunnel the sprayed concrete is loaded by its own self-weight from the moment it is sprayed. This represents some of the most extreme conditions for sprayed concrete. The degree of compaction is least in the crown due to vertical spraying. The bond to the substrate depends heavily on the preparation of that surface and the early age strength gain of the sprayed concrete, which also controls the adhesion between subsequent layers of sprayed concrete. If one of these properties is inadequate or too thick a layer of sprayed concrete is sprayed, lumps of sprayed concrete will sag or simply fall out of the lining (sloughing). The presence of reinforcement will help to prevent this but this would imply that the sprayed concrete is hanging off the reinforcement – potentially leading to plastic settlement cracking.

Autogenous shrinkage occurs when there is no movement of water to or from the concrete. During hydration water is drawn from the capillary pores. This "self-desiccation" causes the cement matrix to contract. Typical values of autogenous shrinkage are 0.004% after one month, i.e. an order of magnitude smaller than plastic shrinkage (Neville 1995). The magnitude of autogenous shrinkage is likely to be greater in sprayed concrete due to the faster rate of hydration and high cement content.

Drying shrinkage occurs in the hardened cement paste as water is lost to the air.[6] First the water from the larger voids and capillary pores is lost, and this causes no shrinkage. However, when the absorbed water in the hardened cement paste is removed, shrinkage occurs. The constituents of sprayed concrete and its curing mean that sprayed concrete is likely to shrink more than a similar strength cast *in-situ* concrete.

Considering the constituents, drying shrinkage increases primarily with increasing cement content, decreasing quantity of aggregate and decreasing stiffness of the aggregate (Neville 1995). The reasons lie in the increased quantity of hardened cement paste and the decreased restraining effect of aggregate. Most natural aggregate itself does not shrink, but shrinkage does vary considerably depending on which aggregates are used. The actual grading curve has little influence other than indirectly by altering the relative proportions of cement and aggregate (Powers 1959). Water–cement ratio has no direct influence, but increasing the ratio reduces the proportion of aggregate. Cement type generally has little influence on shrinkage, though cements which have low gypsum contents tend to shrink more than normal. More accurately, for each cement there is an optimum gypsum content, which minimises shrinkage (Powers 1959). Low gypsum contents also mean a fast reaction, which produces a different gel structure and porosity. This would suggest that the fast reacting sprayed concrete mixes and especially those made with dry mix "spray cements" will produce sprayed concrete that exhibits high shrinkage and shrinkage cracking. On the other hand, it has been suggested that the calcium sulphoaluminate cements exhibit a similar amount of shrinkage compared to normal sprayed concrete, despite their exceptionally fast strength gain. The fineness of the cement also influences shrinkage, because increasing fineness reduces the number of larger particles that restrain shrinkage (Powers 1959). Silica fume, fly ash and GGBS are known to increase shrinkage and are all often used in sprayed concrete. The use of plasticisers and other water reducing admixtures implies a higher cement content in the mix and hence higher shrinkage, although the admixtures themselves are not believed to cause additional shrinkage. Fibres – particularly micro polypropylene fibres – have been found to reduce shrinkage in sprayed concrete (Morgan et al. 2017).

Considering curing and the tunnel environment, one would expect that any measure that reduces moisture loss from the concrete would reduce drying shrinkage. In the extreme, concrete stored underwater actually swells rather than shrinking. Drying shrinkage increases considerably with decreasing relative humidity. Shrinkage at a relative humidity of 40% can be three times greater than at a relative humidity of 80%. However, concrete is subject to a series of competing influences. For example, prolonged moist curing reduces drying but also reduces the quantity of unhydrated cement available to restrain shrinkage (Neville 1995). Well-cured concrete shrinks faster and, since it is more mature, the capacity for creep is much reduced. This reduces the ability to reduce the stresses due to shrinkage. On the other hand, the more mature concrete is stronger.

The effect of ventilation depends on the rate at which moisture can move within the concrete. During the early stages, increased ventilation may increase shrinkage (Kuwajima 1999). At later ages, the rate of evaporation is much greater than the rate of movement of water in the concrete, and so increased ventilation has much less effect. In the case of tunnels, the movement of air stems from tunnel ventilation and may be highly localised in nature, since the forced ventilation is provided by means of ventilation ducts. Typically, in temperate climates, relative humidity in a tunnel is around 50% (though it may be higher) and the temperature is fairly constant within the range from 12 to 24°C, depending on the time of year. The flow of air from ventilation ducts will dry out the sprayed concrete adjacent to them. Concrete with a temperature of 25°C (in air at 20°C and 50% RH), being dried by a current of air at 10 km/hr (which is 2.8 m/s), would lose around 0.5 kg of water per m² per hour (see Figure 2.20). If there was no flow of air, it would lose around 0.15 kg per m² per hour. Considering Figure 2.20, one can see that given the high local air velocities at the end of

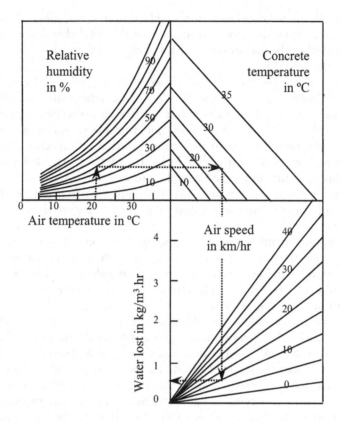

Figure 2.20 Water loss from concrete (after Oberdörfer 1996).

a vent duct (around 20 km/hr), the high temperature of the sprayed concrete during initial hydration (typically 30 to 45°C) and the large surface area, considerable volumes of water could be lost (adjacent to the vent duct) in the early stages of hydration (Oberdörfer 1996). Having said that, the most vulnerable area, at the outlet of the duct, only represents a small proportion of the total surface of the lining. The problem of shrinkage can be more serious in hotter climates.

Furthermore, experiments on shrinkage of sprayed concrete have yielded some contradictory results. Increased ventilation has been found to increase the rate of shrinkage but to reduce the total magnitude (Cornejo-Malm 1995). The magnitude of the shrinkage will depend on the origin of the water that is being removed, i.e. "free" water in capillary pores or "absorbed" water in gel pores, which in turn depends on the original water–cement ratio, the degree of hydration and the porosity of the aggregate (Powers 1959). The effect of shrinkage and curing also depends heavily on the thickness of the layers of concrete (Ansell 2011).

Drying shrinkage continues to take place over years, albeit at a much-reduced rate. Typically, only 20 to 50% of the total shrinkage will have occurred within the first month and about 80% of the total within the first year (Neville 1995).

Curing – the prevention of water loss – is recognised to be important for proper hydration of concrete. While moist curing is often specified for a period of between four to seven days after construction, it is difficult to achieve in tunnels. Covering with impermeable sheets or wet matting is usually deemed impractical in a tunnel during construction. Similarly spraying with water is not preferred by contractors, but arguably it is simple to do. Curing compounds can be applied to external faces or internally as special additives. Externally applied compounds have to be removed before additional layers of concrete are cast or sprayed to ensure that the bond is not impaired.

Carbonation shrinkage occurs in the surface layers of concrete. Carbon dioxide from the air forms carbonic acid, which reacts with various hydrates in the hardened cement paste, notably calcium hydroxide. Hence, this shrinkage is irreversible. The rate of carbonation slows as the depth of carbonation increases, because the carbon dioxide has further to permeate and because the products of carbonation reduce the porosity of the concrete (Blasen 1998). Carbonation is greatest at moderate levels of relative humidity – i.e. 50–75% – since both a lack of water and saturation slow the process. Although the levels of carbon dioxide may be higher than normal in a tunnel due to construction traffic, the fact that carbonation occurs at the same time as drying is liable to reduce the overall contribution of carbonation to shrinkage, because the carbonation will be occurring while the relative humidity is quite high. Typical values for the depth of carbonation are 2 to 3 mm after six months (Oberdörfer 1996) and generally less than 15 mm in older tunnels (e.g. in samples from road tunnels of ages up to 15 years) (Hagelia 2018). Environmental factors may increase this. For example, in road tunnels there is a higher concentration of carbon dioxide in the air.

Temperature effects

Expansion and contraction due to temperature changes occur in tunnel linings during the first few days due to the heat of hydration and subsequent cooling. The thermal coefficient of expansion for cast *in-situ* concrete is largely determined by the coeficients of expansion for the cement and aggregate and their proportions in the mix. Typical values for mature concrete range from 4 to 14×10^{-6} per °C (Neville 1995, ACI 209R 1992), and codes often assume an average value of 10×10^{-6} per °C (DIN 1045 1988, ACI 209R 1992). Similar values have been suggested for sprayed concrete (see Table 2.4) but the coefficient may vary with age. Laplante and Boulay (1994) reported values decreasing from about 21×10^{-6} per °C at 8.4 hours

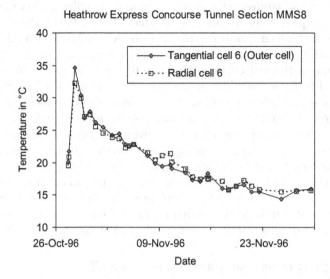

Figure 2.21 Temperature profile in a sprayed concrete lining.

to 12×10^{-6} per °C at 16.4 hours. After that the coefficient remained constant.

Typical profiles of temperature in sprayed concrete linings can be found in Kusterle (1992), Fischnaller (1992) and Hellmich and Mang (1999). Figure 2.21 displays readings from pressure cell temperature transducers and shows how the temperature rises and then decays with time. As one would expect the maximum rise in temperature depends heavily on the thickness of the sprayed concrete layer (see Table 2.8), the initial temperature of the mix and the rate of hydration. The peak rise in temperature occurs about seven to ten hours after spraying for dry mix sprayed concrete and slightly later, at 10 to 15 hours, for wet mix sprayed concrete (Cornejo-Malm 1995). Typically, the maximum temperature lies between the centre of the lining and the extrados and ranges from between 28 to 45°C (i.e. 10 to 25°C above the ambient temperature). Fischnaller (1992) suggests that wet mix sprayed concrete produces higher temperature rises while Cornejo-Malm (1995), quoting lower figures, suggested that wet and dry mix produce similar temperature rises. Typically, after 48 hours, the maximum

Table 2.8 Maximum temperature rises in sprayed concrete linings (Kusterle 1992)

Thickness of lining in mm	Max. temperature rise in °C
50–100	6–9
100–150	10–15
300	25

temperature rise (above ambient temperature) has fallen to less than 20 to 30% of the peak temperature rise (Kusterle 1992). Given the short-lived nature of the high temperatures, they are not believed to have a detrimental impact on the strength of the concrete itself (see also Section 2.2.1).

Assuming a thermal coefficient of 10×10^{-6} per °C and a rise of 20°C, the heat of hydration would induce a maximum compressive linear strain of 0.02% for a perfectly confined sample of concrete. As the concrete cools, it will contract by 0.02% over the following 48 hours. The initial tendency to expand tends not to induce much compressive stress (because the elastic modulus is still small and creep rates are high), whereas the contraction could induce significant tensile stresses in lightly loaded linings. Although shrinkage of a uniform ring would not induce tensile stresses, sprayed concrete linings are made up of a series of panels of different ages. Therefore, there is the potential for differential shrinkage and partial restraint. That said, since the concrete in a lining is not fully restrained so the influence of shrinkage may be small.

Cracking due to shrinkage and temperature effects

The very early ages at which cracking is most likely to occur, the variation in conditions within the tunnel (e.g. ventilation duct in the crown, invert covered with excavated material), the heat of hydration and the possibility of water transfer from the ground all complicate the prediction of cracking due to shrinkage. Indeed, it has been reported that water from the ground can actually lead to swelling in the invert and generally a reduction in shrinkage (Kuwajima 1999). Golser et al. (1989) reported that the shrinkage of the tunnel lining is greatest in the crown, 50% smaller at axis level and negligible in the invert. Creep of tensile stresses (due to shrinkage or bending moments) may lead to additional cracking (Negro et al. 1998), although the case study cited may not be representative of general conditions in tunnels. While the temperature in the thin shell of a tunnel lining peaks and falls much more quickly than, for example, a base slab, which might take several weeks to return to ambient temperature (Eierle & Schikora 1999), the stiffness of the sprayed concrete rises much faster than normal concrete, and so there is just as much risk of the stress induced by the contraction exceeding the tensile strength of the concrete. Also, it has been suggested that the stiffness of concrete rises faster than strength (Eierle & Schikora 1999, Chang 1994). Typical values for shrinkage are 0.10 to 0.12% for wet mixes after 100 days and 0.06 to 0.08% for dry mixes after 180 days (Cornejo-Malm 1995). Given that the concrete is not fully restrained in a tunnel lining, it is the non-uniform nature of the volume change, rather than merely the magnitude of the shrinkage, which causes cracking.

According to some experimental evidence, the general restraint of shrinkage by reinforcement is quite small. The uniaxial shrinkage strain for fibre reinforced sprayed concrete (both polypropylene and steel fibres at low to moderate dosages) was only 8% less than that of ordinary sprayed concrete after 300 hours while 0.39% (by area) steel bar reinforcement reduced the shrinkage strain by 16% (Ding 1998). The addition of fibres may increase porosity (Chang 1994, Ding 1998, Brite Euram 1998), leading to higher shrinkage and creep at higher dosages (60 kg/m³ or more) (Ding 1998).

2.2.7 Creep

Theories and mechanisms

Creep is defined as the increase in strain with time under a sustained stress, and relaxation is the decrease in stress with time in a sample under constant strain (Neville et al. 1983). Relaxation is also sometimes referred to as creep, and here the comments on creep can be taken to apply equally to relaxation unless otherwise stated. In discussions on creep, the term "specific (or unit) creep" is often used. Specific creep is the creep strain per unit stress (typically in units of 10^{-6}/MPa).

Creep can be divided into two components, depending on moisture movement. "Basic creep" is the creep that occurs under conditions of no moisture movement to or from the sample (i.e. conditions of hygral equilibrium). "Drying creep" is the additional creep, which occurs during drying of the sample. The total creep is the sum of these two components. Furthermore, creep components can be divided into reversible and irreversible parts (see Figure 2.22 and England & Illston 1965). On unloading,

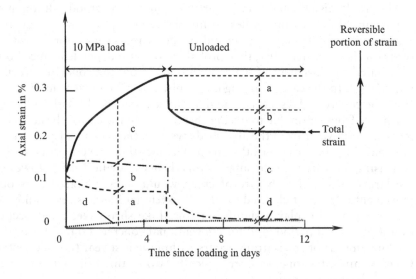

Key:
a Elastic response   c Irreversible creep
b Delayed elastic response   d Shrinkage

Figure 2.22 Decomposition of strains according to the Rate of Flow Method (after Golser et al. 1989).

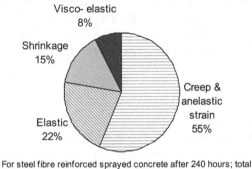

For steel fibre reinforced sprayed concrete after 240 hours; total
strain = 0.293%

Figure 2.23 Composition of strains in a creep test (after Ding 1998).

along with the instantaneous elastic recovery, there will be a gradual
recovery of a portion of the creep. While this is relevant for conditions
of varying stress, it can be ignored if unloading does not occur. In any
case, experimental evidence for sprayed concrete suggests that reversible
(visco-elastic) creep forms a very small percentage of the total strain (typically less than 10%) in samples that are loaded for a prolonged period,
i.e. more than seven days – see Figure 2.23 (Huber 1991, Abler 1992,
Fischnaller 1992, Ding 1998, Probst 1999).

The mechanisms behind creep are not fully understood, although it
is recognised that its origin lies within the cement paste. Shrinkage and
creep are normally assumed to be independent, and a simple superposition
of strains is used. In reality, they probably are not independent since both
are related to movement of water within and from the concrete (Neville
et al. 1983). In the case of drying creep, obviously the movement of water
from the concrete plays a role, and in practice it may be difficult to distin-
guish this from strain due to drying shrinkage.[7] In the case of basic creep,
movement of water from the absorbed layers on the cement paste to internal voids may be a cause of the creep. The fact that creep increases with
increasing porosity tends to support this theory (Neville 1995). However,
the largely irreversible nature of creep would suggest that the viscous
movement of gel particles and to a lesser extent (at higher stresses) microcracking may also play a significant role. Like shrinkage, creep occurs
over a prolonged period, and for conventional concrete 60 to 70% of the
final magnitude of creep strain occurs within the first year (Neville 1995).
Creep strains after one year are typically two to three times the magnitude of the elastic strain.[8]

It appears that creep of concrete under uniaxial tension may be 20–30%
higher than in compression, but relatively little work exists on this subject and some of it is contradictory (Neville et al. 1983). No specific work
on creep in tension of sprayed concrete has been found in the course of

researching this book. Creep in tension reduces the risk of cracking due to uneven shrinkage. Creep in compression will reduce the compressive stresses induced by thermal expansion during hydration and therefore increase the risk of tensile stresses forming on cooling.

Lateral creep has the effect of increasing the apparent Poisson's ratio (creep Poisson's ratio) in uniaxial tests, as creep strain occurs in the direction of the lateral expansion, unless the stress is lower than half the strength (Neville 1995). At lower stresses Poisson's ratio is the same as normal (i.e. about 0.20). However, experimental results from uniaxial compression tests on sprayed concrete vary considerably: $\upsilon = 0.11$ to 0.5 (Golser & Kienberger 1997, Rathmair 1997). Even so one can at least say that the effect of creep is an overall decrease in volume. Under multi-axial stress, the apparent Poisson's ratio is normally lower – 0.09 to 0.17 – and considerable creep will occur even under hydrostatic compression (Neville 1995). The simple superposition of the creep strains due to the stress in a given direction and Poisson's ratio effect of the creep strains in the two other normal directions is unlikely to be valid (Neville 1995, Mosser 1993), but any errors may be small in the case of tunnel linings due to the predominately biaxial stress conditions.

Influences on behaviour

Creep of concrete and sprayed concrete alike increases with decreasing **relative humidity** (i.e. increasing drying), increasing **cement content**, increasing **stress** and decreasing **strength** (Neville et al. 1983, Huber 1991, Fischnaller 1992). The latter two explain why under a constant stress, applied at an early age, creep is greater for more slowly hydrating concretes (since the stress–strength ratio will be higher). However, if one considers concretes loaded with the same stress–strength ratio, the creep is lower for more slowly hydrating concretes (since the magnitude of the stress applied is lower). In the case of uniaxial compression, creep is proportional to the applied stress at low stresses (up to 40% of the uniaxial strength (Pöttler 1990, Huber 1991, Aldrian 1991)). Above this level, it is believed to increase at an increasing rate. At very high stresses (> 80% f_{cu} (Abler 1992)), creep will lead eventually to failure (so-called "tertiary" creep – Jaeger & Cook 1979).

Creep and creep rates are significantly higher at an **early age** of loading since the strength is lower. Byfors (1980) found that for plain concrete the creep strain of a sample loaded at ten hours could be 50 times the creep strain when loaded at 28 days. A sample loaded at an age of eight days may creep by 25% more than a similar sample loaded at 28 days (Huber 1991). This is of importance for SCL tunnels, since the lining is loaded from the moment it is formed. On the other hand, sprayed concrete exhibits a rapid development in strength so, after 24 or 48 hours, the creep behaviour is relatively close to that at greater ages (Kuwajima 1999).

Creep is also influenced by the **aggregate** in the mix, since it restrains the creep. Increasing the proportion of aggregate reduces the magnitude of creep, but the effect is small for the ranges of proportion of aggregate in normal mixes. Creep decreases if stiffer aggregates are used.

Other mix parameters (such as water–cement ratio and cement type) appear to influence creep only insofar as they influence strength and its growth with time (Neville et al. 1983). Hence the types of cement or cement replacements used do not themselves appear to affect (basic) creep behaviour, but their effect on the rate of strength gain will influence creep. Cement replacements, which reduce the porosity (such as microsilica) may well reduce drying creep, since they will restrict water movement.

Considering other influences, the creep of concrete increases with temperature (ACI 209R 1992), but for the case of most tunnel linings, since the increase in temperature due to hydration is relatively small and short-lived, this effect can probably be ignored. This may not be the case in deep tunnels where the ambient temperature of the rock is relatively high. Creep decreases with increasing size of the specimen since this affects drying (ACI 209R 1992, Huber 1991). Some experimental evidence has shown that reinforcement (both bars and fibres) reduces creep (see Figure 2.24, Ding 1998, Plizzari & Serna 2018). Presumably this is

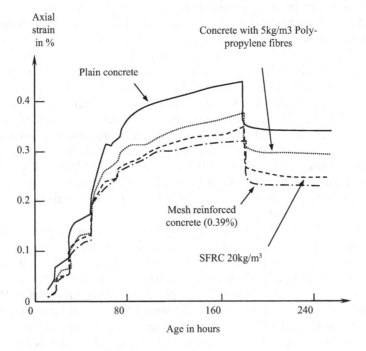

Figure 2.24 Creep test (after Ding 1998).

due to its restraining effect. Typically, 20 kg/m³ of steel fibres (0.21% steel by volume) and 0.39% bar reinforcement reduce the magnitude of creep by the same amount, roughly 25% after 180 hours compared to plain concrete (Ding 1998). Due to their distributed nature, the fibres have more effect than bar reinforcement. Similarly, Plizzari and Serna (2018) report that in one study creep tests on beams with both steel bar and fibre reinforcement performed better than bar reinforcement alone. The macrosynthetic and steel fibres in that study performed equally well. Similarly, in another detailed study no substantial difference was observed between sprayed concrete reinforced steel fibres, macrosynthetic fibres and welded wire mesh during creep loading over one year (Larive et al. 2016). Notwithstanding this, steel fibre reinforced sprayed concrete itself can exhibit considerable creep potential, with creep coefficients of 3 to 6 after one year (depending on the degree of loading) – MacKay and Trottier (2004). Macrosynthetic fibres can have a higher creep capacity, and the creep coefficient for sprayed concrete reinforced with them can be twice as high as that for steel fibre reinforced sprayed concrete (Bernard 2004a, MacKay & Trottier 2004, Bernard 2010). Kaufmann et al. (2012) reported lower values of creep in beam tests with macrosynthetic fibres (i.e. creep coefficients < 3) and a trend towards stabilising deformations, even at utilisations of up to 60%. The relevance of creep in the design depends heavily on the specifics of the tunnel, and the creep performance of fibres depends on many factors, including the concrete matrix, the fibres themselves and the degree of loading.

2.2.8 Variation in properties with environmental conditions

The effects of relative humidity in tunnels on shrinkage and creep and of temperature on strength gain and creep have already been discussed. Air pressure, production influences (including curing and spraying) and the loading on the lining (including exceptional events such as fires) are the other main environmental influences on material behaviour.

Due to several high-profile tunnel fires (Bolton 1999, Bolton & Jones 1999), the **fire resistance** of tunnel linings has come under scrutiny. Sprayed concrete has in the past been used as fire protection in tunnels (Kompen 1990). However, the fire loading in a tunnel tends to be quite severe (Varley & Both 1999). Explosive spalling due to the build-up of moisture within the concrete or differential expansion in the concrete would be likely in most conventional sprayed concrete or segmental tunnel linings unless steps are taken to avoid this. Although less dense than the concrete used for segmental linings, sprayed concrete may have a higher coefficient of thermal expansion and so is unlikely to perform any better than conventional concrete. In the worst case the entire lining thickness could be destroyed. One countermeasure is the addition of small monofilament polypropylene fibres to the concrete mix. During a fire, the polypropylene fibres melt and the resulting capillaries provide an escape path for moisture in the concrete, thus avoiding spalling. Winterberg and Dietze (2004) contains a good review of the state of the art in passive fire protection for sprayed concrete.

Dynamic behaviour is rarely a concern. The seismic design of tunnels is discussed in Section 6.8.1 and the effects of blasting in Section 3.3.2. Since sprayed concrete tunnel linings are not generally subject to cyclic loading, fatigue is not of concern, though one may note that steel fibre reinforced concrete performs significantly better under cyclic loads than plain concrete (Vandewalle et al. 1998). Fibres work well with bar reinforcement under cyclic loading and can be used to reduce the quantity of steel (Bernard 2016). Cyclic loading could be applied to a sprayed concrete lining in a railway tunnel, where it forms the permanent lining and is in intimate contact with the trackbed or is subject to changes in air pressure due to the piston effect as trains pass by at high speed (see the comments on air pressure below).

Sprayed concrete is rarely used in **compressed air tunnelling**. Some evidence exists to suggest that considerable quantities of air are lost through the sprayed concrete lining (Strobl 1991). Research has focused on measuring the air permeability of sprayed concrete linings, so that air losses and supply requirements, as well as surface settlements, can be estimated more accurately (Kammerer & Semprich 1999), and the numerical modelling of construction under compressed air (Hofstetter et al. 1999) – see also Section 6.8.7.

SCL tunnels can also be subjected to variable air pressures due to the piston effect as high speed trains pass through them. This typically imposes a load of ± 1 to 4 kPa, although Holter (2015b) suggested values up to 10 kPa. So long as there is sufficient adhesion of the sprayed concrete to the substrate there should not be any damage to the lining. Since these loads are 100 times less than typical bond strengths so this loading can be discounted in most cases. Otherwise additional measures may be needed to pin the lining back onto the rock (Holmgren 2004).

2.2.9 Durability, construction defects and maintenance

Durability in general

Durability is most relevant for permanent sprayed concrete linings (see Section 4.2.4). In these cases, a design life of 50 to 100 years or more is normally specified. The first point to be made is that sprayed concrete is concrete. For example, the normal standards and guidance for good mix design can and should be applied. The method of compaction is different and the hydration is accelerated but fundamentally it is still concrete.

The durability of the lining is determined by the aggressivity of the environment (both internal and external), the duration of the exposure to degrading actions and its own resistance to those attacks. Considering the sprayed concrete itself, its longevity can be influenced in three areas (after Boniface & Morgan 2009):

- The design of the tunnel (e.g. to minimise early loading and cracking)
- The design of the concrete mix (taking into account the physical and chemical characteristics of each constituent) to counter the demands of the chemical and physical exposure conditions to which the sprayed concrete will be subjected
- The actual physical (and chemical) properties of the *in-situ* sprayed concrete (which can be influenced *inter alia* by workmanship, equipment and quality control)

Design is discussed in Sections 4 and 6, with examples of good specifications for sprayed concrete outlined in Section 6.9, while construction quality control is described in Section 7. Using good quality ingredients in the concrete mix is essential to achieve good durability.

The fundamental questions are:

- Will the sprayed concrete maintain its ability to carry the loads during its design life?
- Will the lining satisfy the desired watertightness during its life?

With this in mind, the specific concerns centre on the stability of the components of hydration (under chemical or physical attack), damage due to the expansion of products (e.g. alkali–silica reaction), the susceptibility of steel reinforcement to corrosion which depends largely on the ingress of harmful chemicals, as well as possible damage to the structure of the concrete due to early loading (see Section 2.2.4). Depending on the waterproofing system (see Section 4.2), the intrinsic permeability of the sprayed concrete may also be relevant. The following sections will cover only those aspects of durability that differ in sprayed concrete, compared to normal concrete.

Maintaining the ability to carry the design loads

Considering the first of these concerns over the long-term stability of the concrete, it is widely believed that the latest additives and accelerators do not have any detrimental effects on the sprayed concrete over the long term (e.g. Myrdal 2011, Hagelia 2018), although the author is not aware of many specific studies on this aspect, using either petrographic examinations or accelerated ageing tests. Melbye (2005) reported that in tests the products of hydration of the accelerated sprayed concrete had been found to be similar to a conventional (durable) concrete. Furthermore, both concretes contained similar patterns of microcracking. Although the sprayed concrete initially showed more microcracking – possibly due to thermal effects – this reduces with time through compressive creep and autogenous healing and the beneficial effects of additions such as microsilica and micro polypropylene fibres.

In terms of sulphate resistance of sprayed concrete, concerns have been raised (Myrdal 2011). Some experimental work has suggested a detrimental impact of non-alkali accelerators (e.g. Spirig 2004) but, nonetheless, the levels of absorbed SO_3 at 100 days, which ranged from 0.69 to 2.10%, were well below the recommended limit of 3.00% (Lukas et al. 1998, Atzwanger 1999). CEM III (sulphate resistant) cements are not generally used for sprayed concrete because they are not very reactive (Garshol 2002). A good mix design using a CEM I cement with a low water–cement ratio and microsilica can produce sprayed concrete with high sulphate resistance (Garshol 2002, Kaufmann et al. 2018).

The normal rules on alkali–silica reaction apply. Some of the additives in sprayed concrete may contain alkalis themselves. Typically, specifications state that, in total, the alkali content of microsilica as Na_2O equivalent should not exceed 2% and accelerators should have an alkali content less than 1% by weight Na_2O equivalent.

Regarding early age loading, ÖBV (1998) cautions that, if sprayed concrete is loaded to more than 80% of its strength, progressive creep will occur and the concrete will be damaged – see also Sections 2.2.4 and 2.2.7. If the lining is designed as part of the permanent works, the safety factors will ensure that loading is well below this level.

In addition to the standard tests involved in mix design, durability can be assessed by examining the permeability of the sprayed concrete (which, typically, must be less than 1×10^{-12} m/s (Watson et al. 1999)), oxygen and chloride diffusion, freeze–thaw resistance, resistance to sulphate attack and the progress of carbonation. Obviously, some of these tests are aimed more at examining the durability of steel embedded in the concrete, rather than the concrete itself. Water permeability may also be assessed by means of a penetration test, according to the German standard DIN 1048 (1991). Penetration depths of less than 50 mm indicate good quality, "impermeable" concrete.

Results from the extensive Brite Euram project suggest that average water permeabilities range from 0.5 to 4.5×10^{-12} m/s, oxygen diffusion coefficients range from 1.79 to 14.2×10^{-9} m/s and chloride diffusion coefficients range from 1.57 to 9.21×10^{-12} m/s. The overall assessment was that sprayed concrete could be produced with as good durability characteristics as a similar conventionally cast concrete (Brite Euram 1998, Norris 1999) – see Figure 2.25. Using samples from real tunnels, Holter and Geving (2016) measured similar values for water permeability, ranging from 0.3 to 33×10^{-12} m/s, with water vapour permeability measurements ranging from 0.7 to 2.2×10^{-12} m/s. Water penetration depths are typically 14 to 25 mm

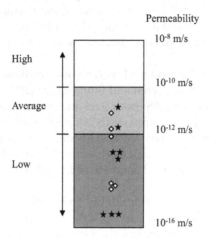

◇ Data from Brite Euram 1998, HEX &
   JLE projects (unpublished)
★ Holter 2015

Figure 2.25 Permeabilities of sprayed concrete vs. categories according to Concrete Society Technical Report 31 (1988).

(i.e. well within the 50 mm limit) (Röthlisberger 1996, Hauck et al. 2011). In a separate study, Yun et al. (2014) reported similar chloride diffusion coefficients, ranging from 4.1 to 20.0×10^{-12} m/s. They also converted the results of rapid chloride permeability tests into diffusion coefficients and found that the results were higher but in a similar range. Furthermore, they confirmed that cement replacements (except for fly ash) improved the chloride diffusion coefficient considerably. In any case, the base concrete admixtures should be free of chlorides such that the percentage of chlorides shall not exceed the limits in standards (e.g. EN 206 which states typically 0.2 to 0.4%).

Similarly, depths of carbonation have been found to be satisfactory – typically, 2 to 3 mm after six months (Oberdörfer 1996) and generally less than 15 mm in older tunnels (e.g. in samples from road tunnels of ages up to 15 years) (Nordstrom 2016, Hagelia 2018).

However, one may note that the test samples are often stored and cured in more favourable conditions than are present in tunnels. Curing measures are rarely implemented on site, because they would slow the advance of the tunnel. Consequently, it is believed that the quality of the sprayed concrete suffers. That said, experimental evidence suggests that the detrimental impacts on the sprayed concrete of less favourable curing conditions and drying are limited (Hefti 1988, Cornejo-Malm 1995, Oberdörfer 1996, Bernard & Clements 2001). The extent of shrinkage and the benefits of curing depend heavily on the lining thickness, especially for thin layers (Ansell 2011). Therefore, the benefits of improving curing may not justify the additional cost and disruption. The addition of additives, such as microsilica, may offer a more cost-effective means of improving durability characteristics.

Sprayed concrete can be produced with a good resistance to freeze–thaw damage – see Section 6.8.8. Myren and Bjontegaard (2014) reported results of scaling tests in which the FRS (both steel and macrosynthetic fibres) performed poorly, despite having apparently acceptable air content (i.e. > 3%) and very high strengths (i.e. > 65 MPa). This may have been due to the use of a solution of 3% NaCl instead of pure water. Nevertheless, they restated the view that in practice sprayed concrete in tunnels has a good resistance to freeze–thaw damage.

In terms of evidence of good durability from real tunnels, Norwegian road tunnels represent one of the best sources of data, given the long history of permanent sprayed concrete and the extensive investigations. Hagelia (2018) presented an excellent summary of these findings for "modern" tunnels (i.e. of ages less than 25 years old), within the framework of the Eurocode exposure classes. He offered recommendations for the minimum thicknesses of fibre reinforced sprayed concrete for each class, whilst recognising the facts that the data only cover a quarter of the typical design life and the rock conditions in Norway are good so most linings are lightly loaded. Overall, the sprayed concrete linings had performed well, even in some chemical aggressive cases (e.g. biofilm attack in subsea tunnels or acid attack due to alum shale).

Durability of reinforcement

As noted earlier, layers of bars and mesh reinforcement are often used in SCL tunnels. Because the sprayed concrete must be sprayed through the mesh, complete encasement is difficult to achieve (Podjadtke 1998). Sprayed concrete rebounds off the bars and "shadows" are left behind the individual bars (see Figure 2.9). Not only does this reduce the bonded length of the mesh but it also provides an ideal location for corrosion of the steel to occur, if water permeates through the lining. Good encasement can be achieved for diameters of 16 mm or smaller (Fischer & Hofmann 2015) – see also section 3.4.4. The spacing of bars laterally seems to be less critical, but the minimum practical spacing is about 100 mm. Alternatively, glass fibre reinforced polymer (GFRP) bars could be used as a non-corroding option.

Steel fibres are often cited as the cure for corrosion concerns because, unlike bar reinforcement, there is no risk of shadowing. Nevertheless, there is still a lesser risk, namely that, where the concrete cracks, the fibres bridging the crack are exposed. If the concrete itself corrodes (e.g. due to sulphate attack), fibres can also be exposed to corrosion as a secondary effect (Hagelia 2018). Corrosion of the fibres will not cause spalling, but it does reduce the load capacity. Various researchers have examined cracked samples of steel and structural macrosynthetic fibre reinforced concrete under different exposure conditions, and the conclusion appears to be that so long as the cracks are narrow (i.e. < 0.15 or 0.20 mm in a less aggressive or only slightly aggressive environment and < 0.1 mm in very aggressive environments), the loss of capacity may be small for steel fibres (AFTES 2013, ITAtech 2016, Nordstrom 2016), and, in the narrower cracks, autogenous healing may occur so the loss in capacity may not be significant. Otherwise the loss in capacity can be significant (Kaufmann & Manser 2013).

Furthermore, this reduction in performance could be exacerbated by embrittlement (see Section 2.2.2). In this process, as the strength of the concrete increases, there can be a tendency for the fibres to snap (brittle failure) rather than being pulled out (ductile failure) (Bernard 2004b, Bjontegaard et al. 2014) – unless high strength steel (e.g. $f_y > 1,500$ N/mm^2) is used (ITAtech 2016) or the concrete mix is optimised. Macrosynthetic fibres perform better and show little signs of deterioration (Bernard 2004b, Kaufmann & Manser 2013, Bjontegaard et al. 2014, Bjontegaard et al. 2018).

As noted earlier, based on experiences in Norwegian tunnels, the minimum thickness for steel fibre reinforced permanent sprayed concrete linings has been set at least 80 mm for benign conditions and 100 mm for aggressive environments in that country (see Hagelia 2018 and Table 2.9). More generally, crack widths can be limited in the design by limiting the tensile stresses in the lining. Cracks which are less than 0.3 mm can close over time due to autogenous healing, and it is worth bearing in mind that cracks in fibre reinforced concrete tend to be more tortuous than in bar reinforced concrete (so-called "crack branching"), which reduces the risk further. In general, fibre reinforcement reduces the permeability of cracked concrete (ACI 544.5R-10 2010).

Watertightness

As noted in one of the preceding subsections, sprayed concrete can be produced with a low permeability (a permeability of $< 5 \times 10^{-12}$ m/s or a water

Table 2.9 Recommended minimum thickness of fibre reinforced sprayed concrete for permanent rock support (Hagelia 2018)

Environment and exposure class according to Eurocodes			
Freshwater	Mildly acidic	Alum shale rock	Subsea tunnels
XC2–XC4	XC 2	XC2–XC4	XC2–XC4
XD1–XD3	XD0	XD0–XD3	XD0–XD3
X0	XA1	XSA	XS3
			XA2–XA3
80 mm	80–100 mm	100–150 mm	100 mm

penetration depth of <25 mm), which is as good as cast concrete – see Figure 2.25. The permeability of the sprayed concrete itself may or may not be relevant, depending on the overall waterproofing concept. This is discussed in more depth later (see Section 4.2.3).

Construction defects

As already mentioned, sprayed concrete and in particular dry mix sprayed concrete is susceptible to poor workmanship. This manifests itself as areas which are too thin or areas of low strength material. Common problems include the failure to clean the substrate, the inclusion of rebound,[9] voids, shadowing behind bars and an intermittent flow of shotcrete (leading to a film of pure accelerator being sprayed on the substrate). The manner in which it is sprayed as well as the quantities of accelerator and water added at the nozzle has a strong influence on the quality of the sprayed concrete. Sprayed concrete is therefore inherently more variable as a material than conventionally cast, ready-mixed concrete. Typically, the standard deviation in 28-day compressive strengths might be 5 MPa for a 35 MPa mix (Brite Euram 1998 – from field trials, Bonapace 1997). This would give a rating of Fair to Poor according to ACI 214-77 (Neville 1995). A similar cast *in-situ* concrete typically might have a standard deviation of about 3.5 MPa, which rates as Very Good to Good (Neville 1995). There is evidence that the quality can be worse where the lining is more difficult to form such as at joints (HSE 2000). Having said that, the use of skilled workers and training can mitigate the risks of poor workmanship, and there are many examples of high-quality sprayed concrete linings for both temporary and permanent uses. EFNARC has published guidance on spraying technique (EFNARC 1999) as well as offering a certification scheme for nozzlemen – see also Section 3.4.4, ACI 506.2-13 (2013) and ACI 506R-16 (2016).

Maintenance

Generally, sprayed concrete linings are designed to function without the need for maintenance. Repair methods are simple and follow the techniques used for conventional concrete. For example, isolated cracks can be sealed using chemicals, if this is needed. Larger areas of defective concrete can be removed (e.g. milled out with a road header or removed by high pressure water), the surface prepared and a new layer of concrete sprayed.

NOTES

1. NB: these formulae often underpredict the very early age strength so a manual correction may be needed (e.g. for ages less than six hours).
2. A more detailed discussion of the hydration of cement and its chemistry can be found in Neville (1995).

3. Bhewa et al. (2018) noted that the difference rises with increasing age from 10% at 28 days to about 20% at 90 days.
4. It should be noted that the formula which Galobarde et al. (2014) proposed appears to overestimate the stiffness at early ages (i.e. less than 24 hours).
5. In this study the concrete contains 6 kg/m³ of macrosynthetic fibres.
6. Plastic shrinkage (see above) is the early part of drying shrinkage while the concrete is still plastic.
7. Creep is time-dependent strain due to applied load; shrinkage is time-dependent strain independent of the applied load.
8. As estimated from Figure 7.1, BS8110 Part 2 (1985), assuming RH = 50%, age at loading = 1 day and an effective section depth of 300 mm. Eurocode 2 (2004) Figure 3.1 suggests large values of about 3 to 5.
9. Rebound is loose material – usually predominantly the larger pieces of aggregate and fibres – which fails to adhere during spraying and falls down.

Chapter 3

Construction methods

This chapter briefly outlines the typical construction methods and plant used to build SCL tunnels. Further details about tunnelling can be found in established texts (e.g. Chapman et al. 2017, BTS 2004, Hoek & Brown 1980). Typical support measures are also described. The excavation sequence is as important in supporting the ground as the sprayed concrete lining or any other additional measures. Hence, as a precursor to designing the sprayed concrete lining, we must first understand how the sprayed concrete functions as part of the support for the tunnel.

Broadly speaking, the ground beneath our feet falls into one of three categories: soft ground, blocky rock or hard rock – see Table 3.1. Obviously, this is simplistic categorisation, and the borders between each class are not clearly defined. Nonetheless this classification is useful because each type of ground behaves in a different way and knowledge of the mode of behaviour guides the design and construction of the tunnel, including the sprayed concrete lining.

To explain the difference in behaviour of the ground we could think of the ground as being like cheese. Soft ground is like soft cheese, for example, brie. It deforms as one body – a continuum. Because of the weak strength of the cheese, it deforms plastically when loaded. Similarly, massive hard rock, like a hard cheese such as Parmesan or a mature cheddar behaves like a continuum when cut or loaded. Deformation is often purely elastic due to the high strength. In between these two extremes are weak or moderately strong, blocky rocks – Wensleydale or Stilton being the equivalent cheeses. Deformation is governed by failure on pre-existing lines of weakness (e.g. joints). In other words, the material behaves as a collection of discrete bodies – a discontinuum.

3.1 SOFT GROUND

Soft ground is defined as soil or weak rock that generally behaves as a single mass (a continuum). Weak rocks can include chalk, breccia or conglomerate. The strength of soft ground ranges approximately from 0 to 10 MPa.

Table 3.1 Types of ground

Category	Soft ground	Blocky rock	Hard rock
Description	Soils and weak rocks	Weak to moderately strong rocks	Massive strong rocks
Mode of behaviour	Continuum	Discontinuum	Continuum
Strengths	< 1 MPa	$1 \ll 50$ MPa	50 MPa $\leq$
Stress–strength ratio*	≤ 1	≤ 1	$\ll 1$
Examples	Sands, clays, chalk	Limestone, sandstone	Basalt, granite

* NB: this is indicative only since the stress conditions are governed by external factors and any rock type can be subjected to overstressing.

The ground requires full support immediately or within a short space of time (i.e. when unsupported it has a stand-up time of less than a few hours). The key mechanisms of behaviour are listed in Table 3.2.

Control of deformations is critical to the success of tunnelling in soft ground – or more precisely, the control of the redistribution of stress is the key. As the tunnel is excavated, there is a redistribution of stress around the opening, and this is accompanied by deformation of the ground. There is often some plastic yielding in the ground. If this process is not controlled, the yielding can lead to excessive deformations or complete collapse. Installation of the tunnel lining – and most importantly the formation of a structural ring – provides support and restricts the deformation of the ground.

3.1.1 Method of excavation

Given the low strength of soft ground, it can be mechanically excavated using bucket excavators (see Figure 3.1) or roadheaders. Tunnel boring machines (TBMs) are often used in soft ground but rarely in combination

Table 3.2 Key mechanisms of behaviour in soft ground

Behaviour	Type of ground	Support measure
Plastic yielding	Clays, sands, weak rocks	Early ring closure; subdivision of face; sprayed concrete
Ravelling	Sands	Ground treatment; subdivision of face; forepoling; sprayed concrete
Heave (of tunnel invert)	Clays, sands, weak rocks and/or high water pressure	Early ring closure
Running	Sands – loose or under the water table	Ground treatment; subdivision of face; forepoling; compressed air; sprayed concrete
Block failure	Stiff jointed clays and weak rocks	Spiling; rock bolting; sprayed concrete

Figure 3.1 Excavation of an SCL tunnel in soft ground.

with sprayed concrete. In soft ground, the speed of advance of TBMs advance normally outstrips the ability of sprayed concrete to be applied and to support the ground.

3.1.2 Support and excavation sequences

Support measures have been developed to cope with the mechanisms of behaviour listed in Table 3.2. The role of sprayed concrete is to provide immediate/short-term support to the ground as a whole, as well as providing support in the long term.

Typically, the lining will consist of 150 to 350 mm of sprayed concrete reinforced with one or two layers of wire mesh or with fibre reinforcement, depending on the size of tunnel and the loads on it (see Figure 3.2 for typical details). Lattice girders are often used to control the shape of the tunnel and to support the mesh during spraying.

If a tunnel cannot be safely constructed using full-face excavation, the face is subdivided into smaller headings. A variety of excavation sequences are used in soft ground (see Figure 3.3). Sometimes they are grouped in the categories of "horizontal division" and "vertical division" of the face. Side galleries are an example of the former, and a top heading drive followed later by a bench and/ or invert is an example of the latter. The choice of excavation sequence is driven by the stability of the unsupported face and heading, but it is also influenced by other factors such as programme and the equipment available. Determining the loads on the sub-headings of a tunnel and the stresses in the linings is complex and will be discussed in more detail later (see Chapter 4). Collapses of SCL tunnels have taught us that these intermediate stages of construction require a full

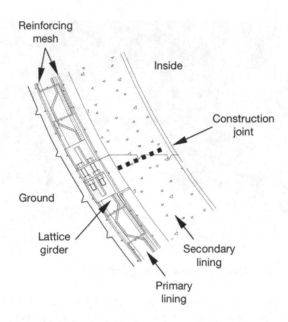

Figure 3.2 Lining details.

design as they may be more critical than the load cases acting upon the finished tunnel.

If the lining is loaded, it will generally continue to deform until a closed structural ring is formed – so-called "ring closure". This applies equally to the temporary headings of the tunnel and the whole tunnel itself. Early ring closure is often specified in designs to control ground movement and settlement of adjacent structures. As an example, Ruzicka et al. (2007) found that a top heading drive ("vertical division" of the face) resulted in more than twice as much movement as a side gallery and enlargement sequence ("horizontal division") due to the difference in distance to ring closure. If plastic yielding is limited to the area under the footings of the top heading, other measures can be used to increase the bearing capacity of the footings. The most simple is enlarging the footing to form an "elephant's foot". The most complex may involve temporary inverts or mini-piles.

Subdivision of the face introduces joints into the lining. For structural integrity, there must be continuity of the steel reinforcement across these joints. Traditionally this was achieved using complex arrangements of lapping bars (see Figure 6.13). It is difficult to build these joints without damaging the lap bars in the process and without trapping rebound when spraying. As a result, the quality of the joints was sometimes poor – both in terms of structural capacity and watertightness. Prefabricated starter bar units, such as KWIK-A-STRIP, have simplified these joints (see Figure 6.14).

(a) Top heading, bench and invert   (b) Side gallery and enlargement

(c) Pilot and enlargement   (d) Twin side gallery and central core

Figure 3.3 Excavation sequences in soft ground.

3.1.3 Special cases

Shafts

The same principles for tunnels outlined above apply to shafts in soft ground (see Figure 3.4). Often the shaft cannot be excavated full face, and to maintain stability, it is divided into sections. Each section is excavated and supported sequentially. Additional measures such as dewatering may be required to ensure the stability of the invert. SCL construction can be used in conjunction with other methods. For example, if the ground consists of water-bearing deposits overlying impermeable ground, the shaft can be started using caisson sinking and continued as SCL after the caisson has been toed into the impermeable layer (e.g. Audsley et al. 1999).

Figure 3.4 Excavation of SCL shaft.

The practical limitation for the gradient of an inclined tunnel is the ability of equipment to move up and down the shaft. Tunnels have been driven downhill at gradients of up to 32% (18.5°) (ITC 2006), while some of the escalator shafts for the Crossrail project in the UK were driven uphill at 30° (Sillerico Matya et al. 2018).

Junctions

At the junction between tunnels or tunnels and a shaft, the lining is reinforced to cope with the redistribution of lining stresses around the openings (see Figure 3.5). As the "parent" tunnel is constructed, it is normal to form the tunnel eye in preparation for the construction of the "child" tunnel. The tunnel eye is where the second tunnel meets the first, and the lining here is thinner and/or lightly reinforced as it will be broken out later. Additional reinforcement is placed around the eye in the shape of a circle, square or bands (see Figure 6.5).

The choice of construction sequence is important at junctions because, when breaking out of (or into) the parent tunnel, the loaded lining is cut and those loads must be redistributed to the adjacent areas. This will cause the tunnels to deform until a complete structural ring is formed again. Hence subdivision of the face, temporary propping and early ring closure are often

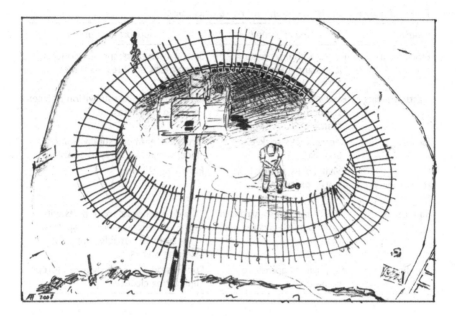

Figure 3.5 Reinforcement around a tunnel junction.

used to manage the stress redistribution and keep the tunnels both stable and within the required limits of deformation.

3.2 BLOCKY ROCK

Blocky rock is defined as rock that generally behaves as a collection of discrete blocks (a discontinuum). Blocky rocks can include rock types such as limestones, sandstones and mudstones. The strength of blocky rock ranges approximately from 10 to 50 MPa, but it may be much higher. The defining characteristic is joint spacing, which ranges from extremely close to wide (< 20 mm to 2 m). The ground requires full support within a short space of time (i.e. stand-up time for the unsupported excavation ranges from a few hours to a few days). The key mechanisms of behaviour are listed in Table 3.3.

3.2.1 Method of excavation

Depending on the strength of blocky rock and type of jointing, it can either be mechanically excavated using roadheaders (see Figure 3.6) or by drill and blast. Tunnel boring machines (TBMs) are often used for longer tunnels. Overbreak along the lines of joints often occurs, leading to an irregular shape of the tunnel surface. Sprayed concrete can be used to backfill this overbreak.

Table 3.3 Key mechanisms of behaviour in blocky rock

Behaviour	Type of ground	Support measure
Block failure	Rocks with intersecting joints and a weak bond across the joint	Spiling; rock bolts; sprayed concrete with mesh or fibres; steel arches
Plastic yielding	Rocks where the stress–strength ratio is greater than 1	Early ring closure; subdivision of face; sprayed concrete
Ravelling	Loose blocky rock, e.g. breccia	Ground treatment; subdivision of face; forepoling; sprayed concrete
Heave (of tunnel invert)	Weak rocks and/or high water pressure	Early ring closure
Squeezing	Where the stress–strength ratio is much greater than 1	"Yielding support": rock bolts with sprayed concrete; slots in sprayed concrete; deformable supports; yielding arches
Swelling	e.g. phyllites, anhydrite, some clays	Subdivision of face; heavily reinforced lining; ring closure

3.2.2 Support and excavation sequences

Support measures have been developed to cope with the mechanisms of behaviour listed in Table 3.3. The role of sprayed concrete is to provide support to blocks and to prevent deterioration of the rock mass over time. The sprayed concrete is often used in combination with rock bolts. Typically, the layer of sprayed concrete will consist of 50 to 150 mm of sprayed concrete reinforced with fibres or wire mesh, depending on the loads on it. The

Figure 3.6 Excavation of an SCL tunnel using a roadheader.

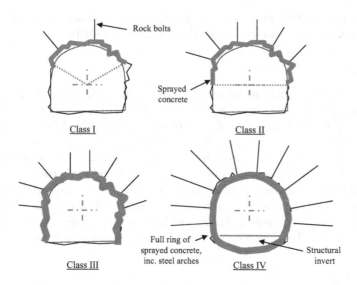

Figure 3.7 Rock support classes.

majority of the *in-situ* ground stresses are redistributed around the tunnel and carried within the rock itself.

Because the behaviour of the ground is governed primarily by the movement of individual blocks, a full ring of sprayed concrete is not normally needed (see Figure 3.7). Support classes III and IV in Figure 3.7 refer to ground that is heavily jointed and is behaving more like soft ground. Again because of the greater stability of blocky rock, compared to soft ground, less restrictive excavation sequences can be used. For example, support does not necessarily have to be installed immediately behind the face, and top heading drives can advance far ahead of the rest of the tunnel. However, the boundary between soft ground and blocky rock is not clear cut. Care must be taken to ensure that stability is always maintained. A common cause of tunnel failures is that a top heading drive is advanced too far in unstable ground.

3.2.3 Special cases

Shafts and junctions

Depending on the ground conditions, many of the comments in Section 3.1.3 may apply. Blocky rock tends to be stronger, and so the support required is less and also it is tailored to controlling block stability. For example, spiles may be installed and heavy thickening of the lining around the opening may not be needed. Longer advance lengths and less restrictive excavation sequences may be employed but only if the stability of the permits.

At junctions there is more scope to reinforce the surrounding ground (e.g. with rock bolts) and to let it carry the redistributed loads.

Swelling

Certain minerals such as phyllite and anhydrite swell when the confining stress is relieved and/or in the presence of water. This can impose enormous loads on a tunnel lining (e.g. Wittke-Gattermann 1998). Apart from minimising the exposure to water, there is little that can be done to prevent swelling. Instead, a heavily reinforced final lining is installed. The best one can do is to manage the convergence of the tunnel and thereby return the situation to equilibrium. During construction, the primary lining may suffer damage and large deformations (tens of centimetres or more) may occur. "Ductile" support may be required, including yielding arches or supports and slots in the lining – see the section on squeezing ground for more details.

Squeezing

Squeezing ground is essentially another form of plastic yielding. Like rockburst, it occurs where the *in-situ* rock stresses are very high. When the uniaxial compressive strength of the rock is less than 30% of the *in-situ* stress, severe or extreme squeezing can occur (see Hoek & Marinos (2000) for a method of estimating the squeezing potential). The NATM design philosophy was originally conceived in part to deal with squeezing ground. It is not economic to install a lining to resist the ground stresses, rather it is better to manage the redistribution of the stresses in the ground as it yields and to install a lining when most of the stresses (acting radially) have relaxed. The deformations that occur before installing a final lining can be up to one metre or more. Rabcewicz (1969) identified that sprayed concrete in combination with rock bolts was very effective in this because it provides ductile support to the ground as it deforms. The rock bolts should be long enough to reinforce the whole of the plastic zone that develops around the tunnel. However, the primary lining can be effectively destroyed in this process. Over the years this approach has been developed further. Longitudinal slots can be left open in the tunnel lining to accommodate the deformations without excessive damage to the sprayed concrete (e.g. the Arlberg tunnel in Austria (John 1978)). The slots can either be left open, or deformable steel supports can be installed in them (e.g. Aggistalis et al. 2004). Grov (2011) describes the use of "reinforced ribs" in squeezing ground. These ribs consist of rings of closed space reinforcing bars, encased in the sprayed concrete.

Creeping

Rock salt, chalk, coal and certain other rocks may exhibit creep behaviour. They will continue to deform when under load. This undermines the arching of stresses in the ground and exerts additional load on the tunnel lining. A full structural ring has to be installed and reinforced to resist these creep loads.

3.3 HARD ROCK

Hard rock covers massive strong rocks that mainly behave as a continuous mass. The joints are widely to extremely widely spaced. The strength is usually greater than 50 MPa and the stress–strength ratio is less than one. While block failure and plastic deformation may occur, in general the rock responds elastically to the excavation of the tunnel. Stand-up time for the unsupported excavation as a whole ranges from days to years. However, individual blocks may need immediate support. The key mechanisms of behaviour are listed in Table 3.4.

3.3.1 Method of excavation

For long tunnels with a constant shape tunnel boring machines (TBMs) are increasingly preferred (see Figure 3.8), while, for shorter tunnels or underground works with a complex shape, the drill and blast method remains the more economic option (see Figure 3.9). Support measures can be installed near the face of the TBM or after mucking and scaling in a drill and blast tunnel. However, if the support is installed immediately behind the face it may risk being damaged by the next blast (see Section 3.3.2).

Depending on rock conditions and the blasting method, the surface of the tunnel can be very irregular in a drill and blast tunnel. Additional sprayed concrete may be required as a smoothing layer to fill up the overbreak.

Table 3.4 Key mechanisms of behaviour in hard rock

Behaviour	Type of ground	Support measure
Plastic yielding	Rocks where the stress–strength ratio is greater than 1	Early ring closure; subdivision of face; sprayed concrete
Block failure	Rocks with intersecting joints and a weak bond across the joint	Spiling; rock bolts; sprayed concrete
Rockburst	Where the stress–strength ratio is much greater than 1	Rock bolts with mesh or SFRS
Squeezing	Where the stress–strength ratio is much greater than 1	"Yielding support": rock bolts with sprayed concrete; slots in sprayed concrete; deformable supports; yielding arches

Spraying robot
mounted here

Figure 3.8 Hard rock tunnel boring machine.

Figure 3.9 Excavation of an SCL tunnel using the drill and blast method.

3.3.2 Support and excavation sequences

The role of sprayed concrete is primarily to support isolated blocks, rather than the rock mass as a whole. The rock itself is often stronger than the sprayed concrete. If the rock is very jointed, then it behaves more like a discontinuous mass and the support is as described in Section 3.2. The sprayed concrete is often used in conjunction with rock bolts. Fibres are tending to replace mesh as reinforcement due to the ease of application and savings in time. There is a risk of the sprayed concrete being damaged by the blasting, but this usually only happens at high particle velocities and early ages. The threshold for damage within the first 24 hours is around 150 mm/s (Geoguide 4 1992). Ellison et al. (2002) suggested that damage can occur at velocities above 500 mm/s but that after 24 hours the sprayed concrete could withstand even this level of vibration. Ansell (2004) and Lamis (2018) have provided more detailed guidance for plain sprayed concrete, based on *in-situ* tests and dynamic numerical modelling of sprayed concrete (see Table 3.5). The time for the adhesion of the concrete to grow to greater than the maximum stress from blasting was found to depend on the thickness of the layer as well as the proximity of the charge and its size. The softer the rock, the lower the stresses in the sprayed concrete. In an extension of this work, Lamis and Ansell (2014) recommended avoiding blasting within 12 hours of spraying (e.g. for a 100 mm thick layer at 3 m from 2.2 kg of explosives in hard, intact rock) which suggests that Table 3.5 is conservative. Numerical modelling has also indicated that current guidelines are conservative (Lamis 2018).

Since relatively small quantities of sprayed concrete are often required in hard rock tunnels, it may more convenient to use pre-bagged dry mix sprayed concrete. This is especially true when the working space is limited such as in small diameter tunnels or shafts or when transport distances are long such as in mines. Thin spray-on liners (also known as thin structural liners (TSL)), which are cementitious polymer-modified spray applied

Table 3.5 Recommended minimum age for sprayed concrete to be subjected to blasting (Ansell 2004)*

Thickness of sprayed concrete (mm)	Distance to charge (m)	Weight of explosives (ammonium nitrate and fuel oil (ANFO)		
		0.5 kg	1.0 kg	2.0 kg
25	4	1 hour	2 hours	4 hours
	2	8 hours	25 hours	65 hours
50	4	3 hours	11 hours	35 hours
	2	45 hours	5 days	9 days
100	4	15 hours	72 hours	7 days
	2	8 days	14 days	Not possible

* See original paper for the additional notes for this table.

membranes, may be an economically viable alternative in these situations (see www.efnarc.org and Archibald & Dirige 2006).

3.3.3 Special cases

Shafts and junctions

Given the stability of the rock, neither shafts nor junctions place particular demands on the sprayed concrete lining. The sprayed concrete is just to control block stability. At junctions, additional reinforcement of the rock mass is required, but this is most effectively achieved using rock bolts. The same applies to pillars between tunnels.

Rockburst

Although the introduction to this category suggested that in hard rock tunnels the stress–strength ratio is less than 1, this may not be the case. In deep mines or tunnels in regions of high tectonic stresses (e.g. the Alps or Himalayas), the rock stresses can exceed the strength. The result is rockburst – the sudden, brittle fracturing of rock around the edge of the excavation. Given the size of the forces involved, rockburst can rarely be prevented, but it can be controlled. Steel fibre reinforced sprayed concrete (SFRS) can absorb a lot of energy as it is deformed. Hence SFRS is often used in combination with rock bolts for protection from rockburst (Bernard et al. 2014), but additional measures may be needed (see Section 6.8.5). A special type of so-called "dynamic" rock bolts have been developed to provide support in these conditions (e.g. Minova's Convergence Bolt).

Fault zones

Within a massive rock mass, there will be major joints or fault zones. Sometimes these are narrow and sealed tight; sometimes they are wide, open and contain water or loose material. These can be very difficult areas because the ground is highly overstressed or the water is at high pressure. There are numerous instances of tunnel collapses at fault zones. In these cases, the ground behaves more like a blocky rock or soft ground, and the support measures described earlier should be applied.

3.4 MODERN SPRAYED CONCRETE

Sprayed concrete technology is a rapidly developing field. While many of the principles remain the same, the equipment has improved greatly in recent years in terms of its ease of use and its capacity. Health and safety considerations and the need for higher production rates are leading to increasing levels of automation. Mix design has also advanced in leaps and bounds. This has been discussed earlier in Section 2.1. Sprayed concrete is produced in two ways: the dry mix process and the wet mix process.

3.4.1 Dry mix sprayed concrete

In the dry mix process, a mixture of naturally moist or oven-dried aggregate, cement and additives is conveyed by compressed air to the nozzle, where the mixing water (and accelerator, if liquid) is added (see Figure 3.10 and Figure 3.11). The dosage of accelerator and the water–cement ratio are controlled by the nozzleman during spraying. In the past, the dry mix has been preferred because it could produce sprayed concrete with higher early strengths, and some countries retain a preference for the dry mix process.

Some of the reasons for choosing dry mix sprayed concrete are listed below:

Dry mix sprayed concrete

- Higher early age strengths (see Table 3.6)
- Lower plant costs
- Small space requirements on site, especially if pre-bagged mixes are used (this is particularly advantageous in urban sites or in mines)
- More flexibility during operation (sprayed concrete can be effectively available "on tap", less cleaning required)

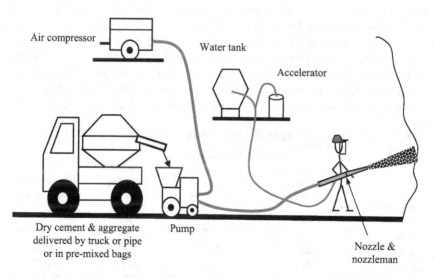

Figure 3.10 Dry mix process.

Figure 3.11 Dry mix pump.

Broadly speaking, dry mix sprayed concrete is best suited to projects that require small and intermittent volumes of sprayed concrete and where there are space constraints on site or long journey times from the point of batching to the face. Dry mix sprayed concrete can be batched and stored in bags ready for use. Accelerator in powder form can be added to the pre-bagged mix so that only water and compressed air are needed when spraying. This simplifies the equipment needed too, but there is no opportunity to vary the dosage of accelerator.

The main disadvantages of the dry mix method are the high levels of dust (see Figure 3.14) and the variability of the product due to the influence of the nozzleman. To counteract this, pre-wetting nozzles (to reduce dust) and

Table 3.6 Compressive strengths of modern mixes (after Lukas et al. 1998)

Age	Dry mix spray cement (oven dry agg.)	Dry mix spray cement (moist aggregate)	Wet mix 6% alkali-free acc.*	Dry mix 6% alkali-free acc.
6 minutes	0.95	0.5	0.5	–
I hour	1.3	1.0	1.0	–
I day	23.0	21.0	15.0	17.0
56 days	41.0	39.0	61.0	39.0

* Higher equivalent cement content in this mix; refer to original report for the full mix details.

Figure 3.12 Wet mix process.

special spray cements, which require no additional accelerator, have been developed (Testor 1997).

3.4.2 Wet mix sprayed concrete

In the wet mix process, ready-mixed (wet) concrete is conveyed by compressed air or pumped to the nozzle, where the liquid accelerator is added (see Figure 3.12 and Figure 3.13). The water–cement ratio is fixed when the concrete is batched outside the tunnel. The dosage of accelerator is controlled by the nozzleman during spraying. There is a global trend towards using the wet mix process in preference to dry mix. The wet mix process is perceived to permit greater control over quality, to be more suited to automation and to be safer (since dust levels are lower). Previously it has been estimated that worldwide about 60% of sprayed concrete is produced by the wet mix method (Brooks 1999), and this share is probably much larger now.

Some of the reasons for choosing wet mix sprayed concrete are listed below:

Wet mix sprayed concrete

- Greater quality control (the mix is batched at a plant and the water–cement ratio cannot be altered by the nozzleman)
- Robotic spraying is required because of the weight of the nozzle and hose, but this leads to higher outputs than dry mix (up to 20 to 25 m³ per hour) and reduces variability due to the human factor. The additional plant cost is partially offset by the reduction in labour costs.
- Lower rebound (typically around 16%, compared to 21 to 37% for dry mixes (Lukas et al. 1998))

Figure 3.13 Wet mix pump.

- Less dust (with dust levels within acceptable limits – see Figure 3.14)
- The use of ready-mix batches, surface scanning and robotic spraying permits records of the exact mix and quantities sprayed to be kept more easily (Davik & Markey 1997), in line with the drive towards using BIM.

Wet mix sprayed concrete is best suited to projects that require regular and large volumes of sprayed concrete and where a batching plant can be located close to the point of use. The cost difference between modern high-quality dry and wet mix sprayed concrete has reduced to the extent that, if one includes all relevant factors (such as rebound, labour costs and cycle time),

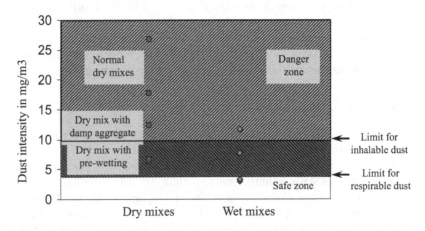

Figure 3.14 Dust levels for different types of sprayed concrete.

Table 3.7 Normalised cost comparisons between dry and wet mix sprayed concrete

Source and notes	Dry mix	Wet mix
Lukas et al. (1998) – all costs exc. rebound	0.72	1.00
Röthlisberger (1996) – all costs	1.12	1.00
Strubreiter (1998) – whole costs over 1,000 m tunnel	0.91	1.00
Melbye (2005) – all costs over 2,000 m tunnel	1.00	0.42

there is little to choose between the two methods (see Table 3.7). On longer tunnels, the higher investment in equipment can be easily justified.

The main disadvantages of wet mix sprayed concrete are the higher plant costs, lower strengths and the limited pot life of the sprayed concrete once mixed. However, the extra plant cost is partially offset by the benefits of automation. As health and safety regulations become increasingly strict, the question of dust levels will strengthen the case for using wet mix sprayed concrete.

3.4.3 Pumping

Large pumps (see Figure 3.13) have two cylinders to provide as smooth a flow as possible. Typical pumping rates are around 8 m³/hr, although the capacity of the pumps can be as high as 20 m³/hr. Before starting to pump concrete, the pump and line should always be washed through with a weak cement grout. Otherwise the water-borne cement in the concrete mix is absorbed as it passes along the line and the first concrete is ruined. If the pump is not operating correctly, the flow will tend to pulse – i.e. the flow is not continuous. This can lead to poor compaction and lamination.

The role of the pump operator is often undervalued. A skilled operative monitors the quality of the concrete entering the pump and the performance of the machine. By doing so he can avert blockages and damage to the pump.

Typical pump lengths are between 20 to 40 m but with the addition of suitable admixtures this can be extended up to 150 m (Melbye 2005). Spirig (2004) describes one extreme example from the Gotthard Tunnel in Switzerland where careful mix design enabled sprayed concrete to be delivered down a vertical pipe up to 800 m deep and sprayed without the use of a remixer.

The length of steel fibres should be less than 50% of the diameter of the pumping hose to avoid blockages.

3.4.4 Spraying

Substrate surface

The ideal substrate for sprayed concrete is a slightly rough, moist surface as this permits a good bond.

Figure 3.15 Control of water ingress.

Concrete cannot be sprayed onto surfaces that it cannot bond to – for example, wet surfaces with running water or smooth plastic surfaces. If water is present, steps must be taken to manage it so that the sprayed concrete can be applied (e.g. see Figure 3.15). Dry mix sprayed concrete can be sprayed onto wet surfaces if the force of the jet of concrete can displace the water and permit the concrete to bond to the ground. Pinning a geotextile over a wet area has the disadvantage of preventing any bond between the sprayed concrete and the ground. As an alternative, drainage holes can be drilled into the ground to concentrate the ingress at discrete points and dry up the surface.

When spraying on to a plastic sheet membrane a layer of thin wire mesh is often placed in front of the mesh, firstly to hold the membrane in place and prevent it from flapping around (so-called "drumming") and secondly to provide something to hold up the sprayed concrete while it is fresh (Jahn 2011).

Spraying technique

The skill of the nozzleman has a great influence on the quality of the sprayed concrete. Several guides exist on best practice for sprayed concrete (e.g. ACI 506R-16 2016 and EFNARC 1999). Virtual reality has been used effectively as a way to train nozzlemen before they go underground (Goransson et al. 2014).

Poor spraying technique can lead to the following defects (see also Figure 3.16):

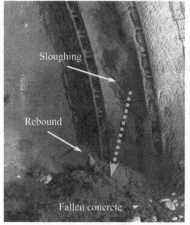
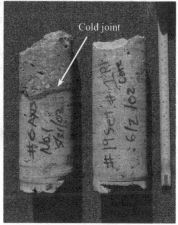

Figure 3.16 Spraying defects.

- Voids: when spraying on an irregular surface, in awkward geometries (such as sharp corners) or around obstructions (such as reinforcing bars), there is a danger of forming voids if the angle of the jet of concrete is wrong.
- Shadowing: voids are formed behind reinforcement bars, exposing the steel to a greater risk of corrosion and reducing the effectiveness of the reinforcement.[1]
- Sloughing: sections of sprayed concrete fall off under their own weight, either because the bond is too weak or because the layer that has been applied is too thick.

- Laminations: rather than being a homogeneous mass, the sprayed concrete may consist of layers with a poor bond between the layers. This may be due to inadequate surface preparation between applications of sprayed concrete or variations in compaction during spraying. White staining may indicate that a film of pure accelerator was sprayed on the surface due to an interruption in the flow of concrete.
- Rebound: if rebound is not cleared away during spraying it may become incorporated into the lining, forming a zone of weakness. Also, excessive rebound is a costly waste of sprayed concrete – see Figure 3.17 for the influences on rebound.
- Low strength: if there is overdosing of the accelerator, there is a risk that the sprayed concrete will have a low strength, either because it has a more porous structure (as the compaction is less effective) or possibly because of a long-term reduction in strength (although this phenomenon does not seem to occur with modern accelerators).

A good design will make the task of the nozzleman easier. Geometries that are awkward to spray should be avoided. Prefabricated starter bar units simplify joints. In practice, it is best to limit the diameter of steel bars to 16 mm or smaller (Fischer & Hofmann 2015). Bars up to 40 mm have been sprayed in successfully, but this is very difficult to do, especially where the bars lap or cross. Multiple layers of smaller bars should be considered as an alternative to individual large diameter bars. The spacing of bars laterally appears to be less critical but the minimum practical spacing is about 100 mm. Spraying overhead is more difficult than spraying at the side of the tunnel (Fischer & Hofmann 2015). The bars should be placed as close to the substrate as possible.

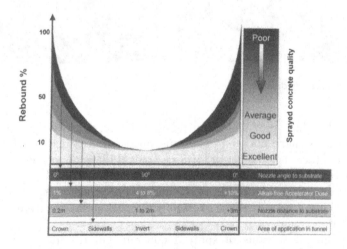

Figure 3.17 Effect on rebound and quality of the principal spraying parameters (Melbye 2005).

The latest technology focuses on reducing the scope for human error. Various distance measuring devices (e.g. laser distometers such as TunnelBeamer[2] or photogrammetric devices such as DIBIT) are in use to check the profiles of the sprayed concrete. One current system that can be used interactively during spraying to check the profile and thickness is the TunnelBeamer (Hilar et al. 2005). Therein lies the strength of this device, outweighing the fact that it can only take spot measurements. The advantage of devices like DIBIT is that they provide a check on the whole surface, but work at the face has to stop for the survey to be done. In a tunnel, the typical accuracy of these systems is ±20 mm which is adequate given that tolerances for spraying are typically ±15 to 25 mm.

Fully automatic (robotic) spraying has been trialled (e.g. for fire protection coatings) but it has yet to be used in a production situation. Undoubtedly this development will come soon.

Nozzles

The nozzle is the device at the end of the hose that converts the stream of concrete (dry or wet) into a jet of sprayed concrete by the addition of compressed air (see Figure 3.18). Nozzles are designed to ensure good mixing of the accelerator and compressed air and, in the case of dry mix, the water too. Special nozzles have been developed for dry mixes to permit pre-wetting of the aggregate just before the main body of water and accelerator is added. This helps to reduce dust.

The nozzle should be cleaned after every use so that it does not become clogged with hardened concrete. The end of the nozzle is designed to blow

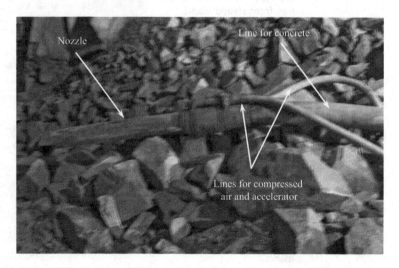

Figure 3.18 Nozzle.

Table 3.8 Finished surfaces of sprayed concrete

Finish	Class acc. to SHW Clause 1708 (HA 2006)	Example
As sprayed – poor quality finish		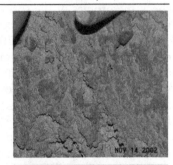
Unfinished	U1 = the concrete shall be levelled and screeded to produce a uniform plain or ridged surface as specified, surplus concrete being struck off by a straight edge immediately after compaction.	
Wood float finish	U2 = after the concrete has hardened sufficiently, it shall be floated by hand or machine sufficient only to produce a uniform surface free from screed marks.	
Steel float finish	U3 = when the moisture film has disappeared and the concrete has hardened sufficiently to prevent laitance from being worked to the surface, it shall be steel-trowelled under firm pressure to produce a dense smooth uniform surface free from trowel marks.	

off, if the nozzle becomes blocked during spraying. Nozzles can suffer greatly from wear and tear due to the abrupt reduction in diameter of the pipe. This can be mitigated by curving the interior of the nozzle or using so-called "stream converter" nozzles (Spirig 2004).

The maximum length of steel fibres should be less than 75% of the diameter of the nozzle to avoid blockages while the maximum length for macro-synthetic fibres is about equal to the nozzle diameter.

Finishing

Depending on the end-user's requirements for the SCL, various finishes can be created (see Table 3.8). The quality of the as-sprayed surface can be improved by reducing the dosage of the accelerator in the final pass when spraying.

The sharp steel fibres protrude from the as-sprayed surface so where people might come into contact with the lining a "smoothing" layer of gunite is often applied as a finishing coat. This smoothing or "regulating" layer is also applied where a spray applied waterproofing membrane will be installed.

Curing

Unlike conventional cast concrete, sprayed concrete is rarely cured. Examples of curing are normally associated with the use of permanent sprayed concrete. A common view – backed up by some research – is that the environment in the tunnel is sufficiently humid (often more than 50% relative humidity) for effective curing to take place (see also Section 2.2.9), although ACI 506R-16 (2016) sets the lower recommended limit for natural curing as 85% relative humidity. This former is a convenient assumption as curing interferes with the construction process in a tunnel. However, it is doubtful that the humidity is this high (Holmgren 2004). On the other hand, if we are aspiring to create concrete that as good as cast concrete, it seems odd that no curing is applied. Curing can be helped by spraying a mist of water onto the sprayed concrete (e.g. Kusterle 1992, Ansell 2011, Grov 2011) or by applying a curing compound to the surface. ACI 506.2-13 (2013) recommends continuous curing for a minimum of seven days or until 70% of the compressive strength is reached, whichever is shorter.

NOTES

1. Research is on-going with a view to proposing a modification to the requirements for bond length in sprayed concrete structures (Basso Trujilllo & Jolin 2018).
2. TunnelBeamer™ is patented by and a registered trademark of Morgan Est and Beton-und Monierbau.

Chapter 4

Design approaches

4.1 DESIGN IN GENERAL

As stated in the preface, this book is not a cook-book and in my opinion, nor should it be. Given the variability of nature and the limitless plausible scenarios for a tunnel, it is impossible to write a simple guide that dictates how to design a tunnel under every imaginable circumstance. The art of tunnelling lies in applying knowledge to the unique situation facing an engineer. To aid this process, this chapter discusses the logic that underpins good design practice.

4.1.1 Observation vs. prediction

In terms of approaches to design, in theory, there is a spectrum from a reliance on pure empiricism to total faith in prediction. In practice neither extreme is used (see Figure 4.1). Given the heterogeneity of the ground, it is impossible to predict in advance precisely how a tunnel will behave, even if the exact construction method is known. Engineers must observe the performance of the tunnel and, if necessary, adjust the excavation sequence and support to suit the prevailing conditions. Similarly, an observational approach does not equate to launching into a tunnel without any concept of what the support will be. Peck's definition of the observational method was succinctly summarised by Everton (1998) as,

> a continuous, managed and integrated process of design, construction control, monitoring and review, which enables previously designed modifications to be incorporated during or after construction as appropriate. All these aspects have to be demonstrably robust. The objective is to achieve greater overall economy without compromising safety.

The terms "robustly engineered" or "fully engineered" design[1] have been introduced to describe a suitable approach to modern SCL tunnelling (e.g. Powell et al. 1997). In a robust design, all the load cases that could be reasonably foreseen should be considered and the support designed to ensure

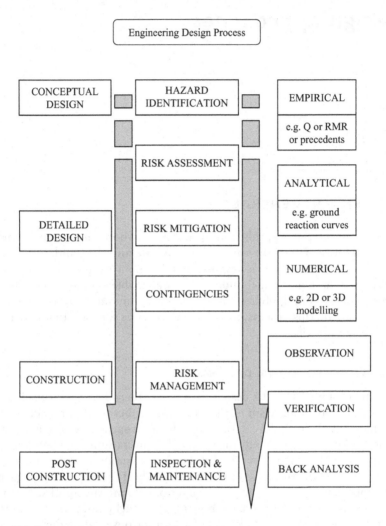

Figure 4.1 Empiricism vs. prediction in design.

that there is an adequate factor of safety, not just for the completed structure but also for all intermediate stages. What distinguishes SCL tunnels from others is the fact that these intermediate stages may be more critical than the final condition.

Designs for SCL tunnels in rock often feature a range of support classes. The choice of support class is based (implicitly or explicitly) on an assessment of which one will manage the prevailing risks best. Even where support classes are not specified in the design, there is usually scope to vary aspects of the design such as advance length and timing of part or all of the support. This should be done within the framework of a set of pre-designed measures. In other words, the key distinguishing feature of modern design practice is that a robust design is produced before the tunnel begins and all changes are made within this framework. A robust design is most easily produced as part of a risk-based approach – see Section 4.1.2.

In rock tunnels under high stresses, the monitoring often feeds into the design in a more active sense. For example, decisions on timing of the placement of the inner lining are sometimes made on the basis of rates of convergence of the lining. If more extreme conditions are encountered than those envisaged, parts of the tunnel may have to be redesigned.

As in SCL tunnels in hard rock, instrumentation is installed in and around a soft ground SCL tunnel to monitor its behaviour. However, its purpose is different. For soft ground tunnels with a robust, fully engineered design, the monitoring is to verify that the tunnels are behaving as predicted rather than to determine the support required. Design changes may still be made, but in general these are within the boundaries of the original design. For example, advance lengths could be varied or optional additional support measures, such as spiling or face dowels, used or omitted, but the thickness of the lining and its reinforcement are not normally altered.

The competence of the designers and the site team are critical to the success of SCL construction, and it is useful to have a designer's representative on site (see Sections 6.9 and 7.4).

4.1.2 Risk-based designs

The UK safety legislation places a duty on designers to identify hazards and avoid or mitigate the associated risks (i.e. reduce them until they are As Low As Reasonably Practical (ALARP)). Many countries have similar regulations. The systematic consideration of risks helps to lead to the production of a "robust" or "fully engineered design" in which all aspects (including temporary cases) are considered in detail before construction begins. These demands complement the needs of SCL tunnels (particularly shallow, soft ground tunnels) since temporary cases are often more critical than the permanent case and the time between the onset of failure and total collapse of a tunnel can be very short. It is this lack of time that means the more observational approach traditionally adopted with the NATM is not appropriate for soft ground.

Risk-based design methods assist all parties to understand how construction, design and safety interact. Inevitably some residual risks will remain, and they are communicated to the contractor via the residual risk register.

4.1.3 General loading

The lining will have to be designed to carry all loads that it is expected to carry (see BTS 2004 for a more detailed discussion). Each tunnel is subjected to its own unique set of loads. In addition to the ground and water loads, these might include compensation grouting, loads from internal sources (e.g. fixtures such as road signs, cable trays or jet fans or objects moving in the tunnel such as vehicles) and accidental loads (e.g. fire or impact from train derailment).

4.2 BASIC PRINCIPLES FOR SCL DESIGN

The fundamental principles that underpin all tunnelling have been outlined in Section 1.2 so only the concluding comment will be repeated here:

> *The art of tunnelling is to maintain as far as possible the inherent strength of the ground so that the amount of load carried by the structure is minimised.*

Good design should incorporate these basic principles, such as *soil–structure interaction* and *"arching"* in the ground. The act of excavating the tunnel modifies the stress distribution in the ground. Figure 4.2 shows one illustration of this, using an analytical solution for a hole in an elastoplastic plate under stress. Introducing a hole into the plate converts a distribution of principle stress in the vertical and horizontal directions into one with high tangential stresses arching around the hole and a radial stress of zero at the edge of the hole. At points far from the hole the stress pattern is unaffected by it. Similarly, by means of arching, a certain amount of the initial stresses is redistributed around the tunnel, leaving the remainder to be borne by the lining (internal pressure, P_i). Hence, deformation of the

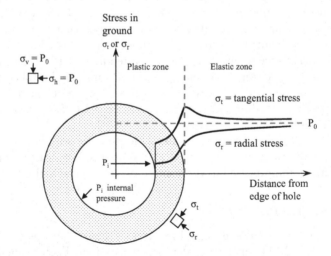

Figure 4.2 "Arching" of stresses around a hole in a stressed plate.

ground is inevitable, and it must be controlled to permit a new state of equilibrium to be reached safely. The arching occurs in three dimensions, and so adjacent excavations may interact. It is important for designers to be able to visualise – even if only in their own mind's eye – their tunnels in three-dimensional form.

The ground can be divided into three categories according to its behaviour:

- Soft ground (soils and weak rocks)
- Blocky rock
- Hard rock

Table 3.2, Table 3.3 and Table 3.4 list the typical modes of behaviour of each type of ground. Soft ground and hard rock behave like a continuum – i.e. as if the ground is a single mass. In soft ground immediate support is required, while in massive hard rock the ground is essentially self-supporting, except for a few isolated blocks. In contrast, blocky rock behaves like a discontinuum – i.e. as if the ground is a collection of discrete blocks. Sprayed concrete is used to support those blocks, often in conjunction with rock bolts.

The fourth dimension – time – plays a major role in tunnelling. The concept of "stand-up" time was recognised long before it was formally articulated by Lauffer and others (see Bieniawski 1984). *Stand-up time* is a measure of how long the ground can stand unsupported. If this is less than the time required to install the sprayed concrete lining, additional measures will be required (e.g. ground improvement). Certain types of ground also exhibit *time-dependent behaviour* such as creep. Time appears as a factor in the support too. Support is usually installed at a finite time after the ground has been excavated, ranging from minutes to days. The support may also be built in several stages. Certain support measures, such as sprayed concrete, display time-dependent behaviour. Time-related aspects of the ground and the support interact due to the progressive nature of tunnelling. For example, each excavation stage throws load onto the part of the tunnel lining nearest the face but, as the face advances, the zone of influence shifts with the tunnel.

This last point highlights another phenomenon, namely that stiffer structures can actually attract load. Installing a heavy lining may not end up being a safer option. Firstly, it is slower to construct, and, secondly, it deforms less and the ground tends to arch preferentially onto the stiff lining. Both can lead to higher loads on the lining.

Finally, all good design should embody the principle of *constructability*. In other words, the structure should be easy and safe to construct. Good constructability will reduce mistakes during construction and improve cost-effectiveness. For this reason, designers should always keep abreast of current construction techniques and innovations.

4.2.1 Ground loads

Due to the phenomenon of arching in the ground, tunnel linings are rarely subjected to the original *in-situ* ground stresses. In most cases, it is overly conservative therefore to design the linings to carry the overburden pressure (i.e. the dead weight of the ground above) or the original *in-situ* horizontal stresses, with the normal partial safety factor for loads applied (e.g. 1.35 according to the Eurocodes). Two more realistic approaches can be used to estimate the ground loads.

Firstly, the loads can be estimated from precedents – e.g. see Table 4.1. The standard partial load factor is then applied to these loads. For soft ground, it is important to note that long-term loads recorded are much less than the full overburden pressure (FOB) and generally well below 60% of FOB (Jones 2018). In contrast, lining loads can be higher for segmental linings (Jones 2018). The same is true of tunnels in rock. That said, unfortunately, there is not an extensive database of information on lining loads in different types of ground for SCL tunnels.

Secondly, where there is inadequate experience of tunnelling particular ground conditions, the ground loads can be estimated from using analytical or numerical methods (see Table 4.2 and also Section 4.3). A relaxation factor can be used (in the convergence confinement method and numerical modelling) to simulate the beneficial effects of arching. The lining is only introduced in the model after the initial stresses have been relaxed according to the factor. Clearly the choice of this factor has a large influence on the final loads on the lining. Table 4.1 includes some typical values. These values are an initial guide only as the amount of relaxation depends on

Table 4.1 Sample long-term ground loads and relaxation factors for shallow tunnels

Ground	Long-term load (as % of full overburden pressure)	Relaxation factor*	Source
London clay	50 to 60%	50%*	Jones 2005
Completely de-composed granite (~ silty sand)	est. 40 to 60%**	75%	Equivalent to volume loss of 0.2%
Conglomerate	–	50%	
Sandstone	–	50–60%	
Chalk Marl	–	50%	Hawley & Pöttler 1991
Chalk	> 50%	25%	Watson et al. 1999
Limestone	–	50–70%	
Rock	See Table 4.2	Up to 100%	

* for more detail on relaxation factors in London clay see Goit et al. 2011.

** although a literature search yielded no data on SCL tunnels, data from pressure cells in segmentally lined tunnels in granular materials suggest values in this range.

Table 4.2 Analytical solutions for estimating ground loads

Ground type	Solution	Source
Soft ground (soils)	Curtis–Muir Wood	Muir Wood 1975, Curtis 1976
Soft ground (soils)	Einstein Schwartz	Einstein & Schwartz 1979
Soft ground (soils and rock)	CCM	Panet & Guenot 1982
Blocky rock	Protodykianov	Szechy 1973
Hard or blocky rock	P arch	Grimstad & Barton 1993

the constitutive model and the construction sequence. More recent work – modelling a multi-stage excavation sequence with a sophisticated strain-softening model for an over-consolidated clay – has shown that several stages of relaxation should be used to replicate better the ground movements and development of load on the tunnel that are observed on site (Goit et al. 2011).

In a similar manner, Muir Wood (1975) recommended that only a fraction of the loads predicted by his analytical solution are applied to the lining. For the specific case of London clay, he proposed a value of 50%.

The experience in high stress rock tunnels and simple analytical models based on stress relaxation have led to the view that the more the ground is permitted to relax, the lower the ultimate load on the lining will be. This has become enshrined in the NATM philosophy. However, this does not appear to hold true for SCL tunnels in soft ground. In fact, the more the ground relaxes the higher the load will be due to strain softening in the ground and the generation of negative pore pressures which later dissipate (Thomas 2003, Jones 2005).

For blocky rock, programmes such as UNWEDGE can be used to calculate sizes of typical blocks, based on the pattern of joints.

Once again, the standard partial load factor is then applied to these loads (e.g. in line with Eurocode 7 Design Approach 2 (EC 7 (2004) clause 2.4.7.3.4.3)). Alternatively, in numerical modelling, partial safety factors can be applied to soil parameters in line with the procedure described in Eurocode 7 Design Approach 1. The numerical model then provides "factored" loads directly. However, this is not recommended as factoring ground parameters may produce a misleading simulation of the ground behaviour. For example, a reduction in shear strength could lead to much more plastic yielding and deformation than might reasonably be expected.

4.2.2 Excavation and support sequence

The way that an SCL tunnel is constructed has a large bearing on final stresses in it and the amount of surface settlement. The age of the material at loading is the main reason for the difference in behaviour of sprayed concrete and conventional concrete in tunnel linings. An increasing but variable load is applied to the sprayed concrete from the moment that it is sprayed. The loading of sprayed concrete at an early age means that creep may be important and the material may be "overstressed" – i.e. loaded to a high percentage of its strength, leading to pronounced nonlinear behaviour and possibly long-term damage to the microstructure. Since the properties of sprayed concrete change considerably with age during the early life of an SCL tunnel, the response of the tunnel lining to loading varies, depending on when the load is applied.

Key elements of the construction sequence are outlined below.

- Excavation sequence: a tunnel is often subdivided into headings. The headings advance in a sequence, with each heading increasing the size of the tunnel until the full section is formed. Typical sequences include: top heading, bench and invert; pilot and enlargement; side gallery and enlargement and twin side galleries and central pillar. The choice of sequence is governed by the overall stability of the temporary headings.
- Subdivision: subdivision of the face of a heading is used to control stability of the face along with measures such as a sealing coat of sprayed concrete.
- Ring closure: the key to controlling ground movements in soft ground tunnelling is the formation of a closed structural ring. While the ring is open, ground movements into the tunnel continue but ring closure brings them to a halt.
- Advance length: also called "round length" in drill and blast tunnels, the advance length is the distance of a heading that is excavated in a single stage.
- Adjacent excavation or construction: because the stresses in the ground arch around a tunnel heading, the adjacent ground experiences an additional load. Any structure in that zone of influence must be designed to cope with that loading. Adjacent construction activities, such as compensation grouting, can also impose loads on a tunnel.

The influences on the choice of excavation and support sequence have been outlined in Section 3. The main ones are listed below:

- Stand-up time (and stability of the ground overall)
- Face stability

In addition, the construction team will have its own preferences on sequence, depending on the construction programme and the equipment available.

The stability of the ground at the face governs the size of the faces in each heading. Various analytical tools can be used to help estimate this, e.g. N – stability number for cohesive materials (Mair 1993); see Leca & Dormieux (1990) for cohesionless ground.

In certain cases, it may be advisable to delay installing support (e.g. swelling or squeezing rock), but in general, and especially in soft ground, it is better to install the tunnel lining sooner rather than later. This will minimise the risk of the ground around the tunnel deforming excessively or weakening.

4.2.3 Water and waterproofing

Groundwater can have two influences on the design of an SCL tunnel. Firstly, it affects the behaviour of the ground, and, secondly, waterproofing measures may be required, depending on the purpose of the tunnel.

Apart from the immediate effect of water on the strength of the ground and its stability, in the longer term, water may also impose loads on the lining. If the tunnel acts as a drain, it may be loaded by seepage forces. In stiff over-consolidated clays, the equalisation of negative pore pressures that have been generated around the tunnel will add load to the lining, although in most cases this increase will be small and actually tend to even out the loading on the lining (Jones 2005).

Table 4.3 shows the ranges of application of the different design approaches for waterproofing.

Table 4.3 Design approaches for waterproofing

Drained	Partially tanked	Fully tanked
Water inflow is acceptable	Minor water inflow is acceptable	Water inflow is unacceptable
Class 1 or 2 * inside final lining Class 3 or 4 * if no internal lining	Class 1 or 2 * inside final lining	Class 1 or 2
$0.02 << 0.1$ or $0.2 << 0.5$ l/m^2 per day Over 10 m	< 0.02 or $0.02 << 0.1$ l/m^2 per day Over 10 m	< 0.02 or $0.02 << 0.1$ l/m^2 per day Over 10 m
Any tunnel under high water pressure Water tunnels	Public areas, metro tunnels, shallow road or rail tunnels	Public areas, metro tunnels, shallow road or rail tunnels
Low or high water pressure	Low to medium water pressure	Low to medium water pressure
< 1 bar or > 5–8 bar	< 5–8 bar	< 5–8 bar
Permeable or impermeable ground	Impermeable ground	Permeable ground
Design for seepage pressures only	The lining may be designed for full water pressure	The lining must be designed for full water pressure
e.g. San Diego MVE – Thomas et al. 2003	e.g. Channel Tunnel UK Crossover – Hawley & Pöttler 1991	e.g. Crossrail – Thomas & Dimmock 2018

The effects of groundwater (either as a direct load or its influence on the stability) must be accounted for in design analyses when determining ground loads (see Section 4.2.1). If the tunnel has an impermeable lining, it will be subject to uplift forces.

A wide variety of measures can be taken to keep water out of a tunnel and the choice depends on the hydrogeology, the purpose of the tunnel and the permissible water ingress. The ITA has produced guidelines on the impermeability of tunnel linings, depending on the use of the tunnel (ITA 1991). Figure 4.3 shows solutions for achieving the categories for dry tunnels such as public areas in stations or heavily used road tunnels. However, as the external water pressure increases (e.g. at around 50 to 80 m of head), it may be more economic to permit a small controlled water ingress, rather than to exclude it completely. This reduces the load on the lining.

Normally, the purpose of the waterproofing is to keep groundwater out, but sometimes it is to keep the fluid in the tunnel from leaking out (e.g. water transfer tunnels). However, it should be noted that the approach to waterproofing also varies widely between different countries and different sectors, depending on the client's budget and local design practice. For example, the limit on water ingress in a subsea road in Norway is 14 l/m^2 per day (over a 10 m length), which is far above ITA Class 5, whereas a similar road tunnel in the UK would have to meet Class 2. The difference arises in part from the lower number of tunnel users but partly also

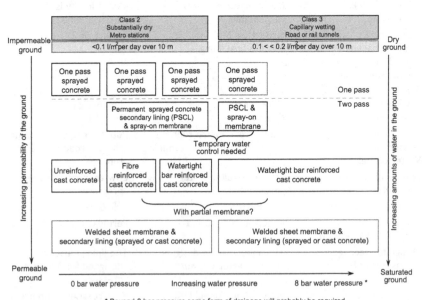

Figure 4.3 Options for "undrained" solution to achieve a dry tunnel.

from national preferences. Waterproofing is a notoriously difficult part of construction, and 100% success cannot be guaranteed even with expensive "fully tanked" membrane systems. One small flaw can compromise the whole system. The impermeability of the sprayed concrete is only one element of the whole waterproofing system for a tunnel. Joints in the lining or waterproofing system itself and between the tunnel and connected structures, such as shafts, represent weak points, and extra care is required to seal them against water ingress.

Waterproofing represents a huge topic in its own right. Further information can be found in the BTS Lining Design Guide (BTS 2004). Some brief notes have been included in the following sections to explain the impact of waterproofing measures on the sprayed concrete lining.

Controlled inflow (via drainage holes)

A "drained" solution is often adopted where complete watertightness is not required (e.g. in non-sensitive tunnels which are not open to the public), where the water ingress is transient and low in volume or where the water pressures are so high that it would not be economic to exclude the water (Thumann et al. 2014).

Figure 4.4 shows a typical arrangement of a drained solution. Small diameter plastic pipes (e.g. 50 mm) with slots cut in them are often inserted into the holes to keep the inflow path open. The pipes can be wrapped in a geotextile to prevent fines from the ground being washed in. The inflowing water is then directed into drains running in the invert of the tunnel. Unfortunately, salts such as calcium hydroxide can be leached out of the

Figure 4.4 Drainage pipes.

ground and the sprayed concrete lining and then deposited in the drains – a process known as "sintering" (Eichler 1994, Thumann et al. 2014). In the worst case, the drains can become completely blocked, creating a significant maintenance task (e.g. in many Alpine tunnels). In some cases, where a drained solution is adopted, the tunnel lining is checked against the load case of a full hydrostatic head (i.e. assuming the drains become blocked) – e.g. Channel Tunnel UK Crossover cavern.

This approach can be combined with "strip drains", sheets of plastic drainage layers which are pinned to the rock with bolts and which direct the water down to drains in the invert. Where the water is seeping through the ground in general or through systematic drainage holes, rather than in isolated places, these drainage layers can be used to cover larger areas. A disadvantage of this form of "waterproofing" is that the thin layer of concrete (~50 mm) sprayed over the drainage layer tends to crack with large cracks due to the unrestrained shrinkage (Ansell et al. 2014). A discussion of the ways to overcome this can be found in Ansell (2011).

Drainage holes can also be used during construction to control inflow. Instead of the water seeping in through the whole lining, it becomes concentrated at the holes, and so it is easier to handle. It may also be possible to draw down the water table locally which reduces the inflow pressure.

Alternatively, the permeability of the ground can be reduced to limit the inflow. This sort of systematic pre-grouting is common in Scandinavia (NFF 2011, Smith 2018), and it can be used to reduce the water inflow to as low as 2 to 10 litres per minute per 100 m of tunnel (Franzen 2005, Grov 2011).

Sheet membranes

In the past, sprayed concrete on its own has not been considered watertight and additional measures have been required. Also, concerns over the durability of sprayed concrete encouraged designers to ignore the primary lining once the secondary lining had been installed. A common solution is to install a sheet membrane inside the primary sprayed concrete lining. In this so-called "fully tanked" or "partially tanked" solution, a secondary lining is installed inside the membrane to carry the water load and normally the entire ground load in the long-term. Even if the primary lining is included in the long-term design, because the membrane introduces a frictionless interface between the two linings, it must be assumed that no composite structural action occurs. Adopting this approach has a major impact on the design of the lining (see Section 4.3.2), but the sheet membrane itself too has an influence on the sprayed concrete lining.

Typically, the surface of an SCL tunnel will have to be prepared before a sheet membrane is installed to prevent it from being punctured (Figure 4.5). Sometimes the drainage fleece can serve the additional purpose of a protective layer. If there are sudden changes in profile or deep depressions in the surface of the lining, there is a risk of over-stretching

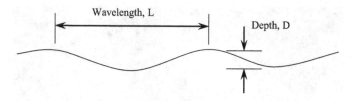

Figure 4.5 Depth to wavelength criterion for smoothness.

the membrane during concreting. To avoid this, smoothness criteria are set – e.g. see Table 4.4.

Various types of waterproofing sheet membranes are available (see BTS 2004), and their installation is a highly specialised job (see Figure 4.6). Temporary drainage measures may be required during the installation before the secondary lining is in place to resist the water pressure.

In a "fully tanked" solution, the membrane is designed to form a completely impermeable barrier to water inflow (or outflow). In a "partially tanked" solution, the membrane does not extend around the full circumference of the tunnel. It terminates at either side of the invert in a drain, and it is combined with a geotextile fleece to transmit the water to the drains. The advantage of this is that the inner lining does not need to be designed to withstand the full hydrostatic pressure. On the downside, measures must be taken to prevent sintering (see the section on controlled inflow).

Even in the "fully tanked" solution, it is prudent to install some drainage in the tunnel as it is almost inevitable that there will be a leak somewhere in the membrane.

Spray applied waterproofing membranes (SAWM)

As a development of the chemically based thin structural liners (TSL), a new technology was introduced in the early 2000s whereby a membrane can be sprayed *in-situ* to form a waterproofing layer (Thomas & Dimmock 2018).

Table 4.4 An example of criteria for smoothness of SCL tunnels

Parameter		Limit
D/L – as shown in Figure 4.5 – (over a 3 m reference length)	Temporary linings	less than 1:5
	Permanent linings	less than 1:20
Transitions and intersections of underground structures shall be rounded off with a minimum internal radius of		500 mm
The minimum thickness of the final finishing layer		25 mm
The radius of curvature of the finishing surface over protruding steel parts such as rock bolts		Greater than 200 mm
All protruding steel shall be cut flush with the surface unless treated with additional shotcrete		

Figure 4.6 Sheet membrane installation.

Full details can be found in an ITAtech guidance document (ITAtech 2013).

These products are usually aimed at providing a more cost-effective solution than sheet membranes in situations with low hydrostatic pressures or complex geometries. A key advantage is the ease of application, using simple equipment that is easy and quick to operate. Depending on the bond across the membrane, it may be possible to use spray-on membranes between layers of a composite lining – see Section 4.3.

A key element of the original SAWM concept is that the membrane is bonded on both sides to the substrate. This offers two advantages: firstly, no migration of water from any open cracks in the primary lining behind the membrane; secondly, composite structural action between the primary and secondary concrete linings and the membrane. This offers the potential for more structurally efficient (and therefore more economic) linings. Generally speaking, SAWMs have been demonstrated to be durable. They are resistant to common environmental influences such as attack from typical groundwater chemistry and frost (down to –5°C). SAWMs have been used for projects with a design life of up to 120 years (e.g. Crossrail (Thomas & Dimmock 2018) – see also the case studies in ITAtech 2013).

While they are impermeable to liquid water, most spray applied membranes are vapour permeable – a bit like Gore-Tex jackets. Hence small quantities of water are transmitted through the membrane (Holter & Geving 2016) and lost to the air inside the tunnel. Holter (2015a) found that the degree of capillary saturation is less than 100% at the membrane, even though the concrete at the interface with the rock is fully saturated.

In theory, this behaviour could permit the use of a reduced value for the pore pressure acting at the membrane in design calculations, for example, perhaps to 85% of the nominal water pressure – as discussed in Thomas & Dimmock (2018).

Good surface preparation is critical to the successful application of a spray applied membrane (ITAtech 2013). As with sheet membranes, particular care is required when forming joints, although, in the case of a spray-on membrane, the joints are simply formed by overlapping the adjacent section onto the previous one. The surface must not be too rough or else there will be a high consumption of the expensive membrane material and flaws (such as pinholes where the membrane fails to bridge across a deep depression in the concrete surface). To avoid this, the sprayed concrete should be a U1 finish or better (see Table 3.8), or a smoothing layer of gunite should be applied. Figure 4.7 shows how the as-sprayed surface roughness can vary depending on maximum size of aggregate in the sprayed concrete.

A disadvantage of spray-on membranes is that – like sprayed concrete itself – the product is created *in-situ* in the tunnel (see Figure 4.8) and hence is vulnerable to environmental influences (such as low temperatures or high humidity which can harm the curing process) and poor workmanship. Also, the curing time of the products, which can last for several weeks, can delay other activities such as installing the inner lining.

4.2.4 Permanent sprayed concrete

Sprayed concrete can be produced with excellent durability characteristics (i.e. equal to that of *in-situ* concrete, as indicated by its permeability and porosity values (Neville 1995, Palermo & Helene 1998, Norris 1999)),

Figure 4.7 Sprayed concrete surfaces covered by MS 345 with different maximum sizes of aggregate (courtesy of BASF).

Figure 4.8 Spray-on membrane installation.

although this increases the unit cost of the material. To be permanent, this sprayed concrete must be durable enough to last for the design life of the tunnel. Sprayed concrete is still not widely used as part of permanent works (at least in public tunnels) (Golser & Kienberger 1997, Watson et al. 1999), and no clear consensus on the performance specification has emerged. In the past, concerns over the durability of sprayed concrete tunnel lining have prevented the more widespread use of the material in the permanent lining of tunnels. The situation was not helped by the fact that the concept of durability is often poorly understood.

There are normally two questions that must be answered before a project accepts the use of permanent sprayed concrete.

1. Is the sprayed concrete durable?
2. Is the lining sufficiently watertight?

Considering the first question, modern good quality sprayed concrete is a durable material (see Section 2.2.9 for a detailed discussion about the durability of sprayed concrete). Poor workmanship, water inflow during construction or excessive loading at an early age are the only significant risks to durability of the concrete.

In general terms the strength should not degrade over time, and the concrete should be dense and have a low permeability for water. The latter criteria are aimed at reducing the potential for water ingress through the body of the lining and corrosion of reinforcement within the lining. Table 4.5 contains typical requirements specified for a "permanent" sprayed concrete

Table 4.5 Typical design requirements for permanent sprayed concrete linings

Parameter	Value	Source/comments
Max. water–binder ratio	0.45	–
Min. cement content	400 kg/m^3	Good to add 5–10% of microsilica (bwc)
Min. compressive strength	Depends on lining loads – typically 30 to 40 MPa	No reduction with time after 28 days
Max. accelerator dosage	Keep as low as possible	–
Water permeability	>= 10^{-12} m/s	Darby & Leggett 1997
Max. water penetration	=< 50 mm	Lukas et al. 1998, EFNARC 2002
Max. crack width	0.4 mm	ÖBV 1998
Curing period	Seven days	By spraying with water
Bond between layers of concrete	1.0 MPa	EFNARC 2002

to achieve those basic criteria. There has been some concern that the loading experienced by the sprayed concrete at an early age may damage its long-term strength. This is more relevant to "single shell" linings (see Section 4.3.4) but, in any case, the normal compressive strength testing of the *in-situ* sprayed concrete should detect any significant damage.

A higher specification is required than for sprayed concrete used in temporary works. Modern specifications typically require compressive strengths at 28 days of 30 MPa or greater (e.g. C35/45 acc. to ÖBV 2013) for permanent sprayed concrete. Table 2.1 contains an example of mix design. Higher standards of workmanship and, hand in hand with that, greater quality control are necessary. Occasionally curing is applied (e.g. Hvalfjordur described in Grov (2011)) but as noted in Section 3 this introduces an additional activity in the tunnel which construction teams prefer to avoid.

Corrosion of reinforcement steel embedded in the sprayed concrete presents the main residual risk. One way to remove this is to use fibre reinforcement. However, there may still be cases such as junctions where heavier, steel bar reinforcement is required. In those cases, good workmanship should ensure that the steel is safely encased in dense concrete with a low permeability – just as in cast concrete (see Section 3.4.4).

Considering the second question of watertightness, as described in Section 4.2.3, a variety of methods can be applied to control water ingress, and many of these are compatible with permanent sprayed concrete, most notably spray applied waterproofing membranes. The options for these types of lining design are discussed in Section 4.3.

Accounts of pioneering projects in the field of permanent sprayed concrete linings can be found in Gebauer (1990), Kusterle (1992), Arnold & Neumann (1995), Darby & Leggett (1997), Zangerle (1998), Palermo & Helene (1998), Grov (2011) and Thomas & Dimmock (2018). A much more comprehensive listing of more than 150 tunnels of all types and from all parts of the world can be found in Franzen et al. (2001).

Clearly installing a secondary lining and ignoring the primary lining costs more, both in time and money, than a lining which uses all the concrete sprayed as part of the permanent lining. Grov (2011) cites one example of 525 m long de-silting chambers on a hydropower project where this change reduced costs by 15% and saved more than eight months' time. Hence modern designs often incorporate permanent sprayed concrete in the permanent works design – see Section 4.3.

4.2.5 Design for fibre reinforced sprayed concrete (FRS)

The section covers the key design aspects to be considered when using fibre reinforcement while the basic properties and behaviour can be found in Section 2.2.2. Fibre reinforcement is increasingly popular for both SCL tunnels in both rock and soft ground. Hence, it is worth elaborating how to design linings with this material. FRS is normally specified on the basis of energy absorption or post-crack ductility. The former is more applicable to rock tunnels while the latter is used more often for soft ground tunnels.

The basic behaviour of FRS has been described earlier but, before going further, it should be noted that the desired "ductile" response of the lining does not automatically mean that requires a deformation hardening response from the FRS (Grov 2011). The vast majority of the tunnels built to date with FRS linings have used deformation-softening FRS. This can be perfectly acceptable because the lining and ground act in concert in the soil–structure interaction. Structurally speaking, this is a highly redundant system which can redistribute stresses within itself as deformation occurs – within the limits of the ground and the lining. One could be forgiven for suspecting that the pressure from suppliers to choose fibres, which offer deformation hardening, is driven as much by commercial preferences as it is by engineering rationale.

In soft ground tunnels, the loads in the linings are typically compared with the capacity based on an axial force–bending moment interaction diagram. Over the years a variety of design methods have been proposed based on stress blocks in the tensile region (Thomas 2014). The most recent and arguably the best is the Model Code 2010 (fib 2010), and, for highly redundant structures like tunnel linings, it can be used for both steel and macro-synthetic fibres – despite its warning about the risk of creep in non-metallic fibres (Plizzari & Serna 2018).

A fundamental problem for the designer with Model Code and other methods is that some of the values required to define the design strengths have to be determined from tests. This can leave designers uncertain about which values can be assumed when no test data is available. Thomas (2014) proposed some reasonable values, based on published strength data, for initial design assumptions when no specific test data is available. These values are $f_{R1k} = 0.5\, f_{fck,fl}$ and grade "b" – i.e. f_{R3}/f_{R1} values between 0.7 and 0.9, as per the Model Code classification (fib 2010). Thomas (2014) noted that the additional capacity added by the fibres tends to be very small, both in absolute terms and in comparison to bar reinforcement. One could almost say that they offer little meaningful increase in bending capacity and one could simply use the interaction diagram for concrete with no reinforcement but safe in the knowledge that the fibres would make the concrete ductile. This would be a somewhat disappointing conclusion but it could avoid arguments over design philosophies and test methods.

For rock tunnels, the performance requirement for the FRS is often given in terms of energy absorption. This is because the (empirical) Q-system chart recommends thicknesses of FRS with certain energy absorptions, ranging from 500 to 1000 J. However, the author is not aware of any analytical design method which uses energy absorption, although one could conceivably develop one based on the virtual work done by a falling wedge. Barratt and McCreath (1995) describe a limit equilibrium analytical design method for sprayed concrete which can be used to design FRS linings (see also Section 6.2.1).

4.3 LINING TYPES

Guided by the design principles above, the design team will choose which type of lining best meets the needs of the project. Table 4.6 lists options for the design of permanent tunnel linings. The key differences between the design cases below arise from the assumptions about the durability of the sprayed concrete and the method of waterproofing (and from that, more specifically, the extent of bonding between the membrane and primary/secondary concrete and where the water pressure acts – see Figure 4.9).

Theoretically, one could have a DSL where the primary sprayed concrete lining is regarded as permanent. Despite the lack of a bond at the interface with the membrane, there would still be some limited load sharing (Thomas & Pickett 2012), but the economic benefits would be so limited that it would not be worth considering.

Table 4.6 Lining types and key characteristics

Lining type	Primary Lining		Waterproofing	Secondary Lining	
	Type	Permanent		Type	Permanent
Double shell lining (DSL)	SCL	No	Sheet	Cast	Yes
Composite shell lining (CSL)	SCL	Yes	Sprayed	SCL	Yes
Partial composite lining (PCL)	SCL	Yes	Sprayed	SCL	Yes
Single shell lining (SSL)	SCL	Yes	none	N/A	

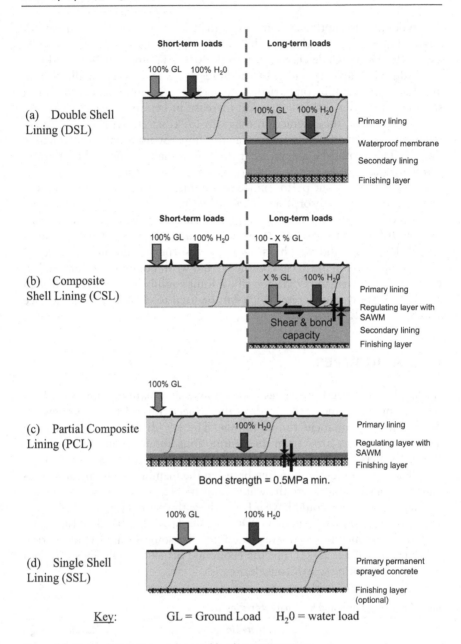

Figure 4.9 Lining design approaches and loading assumptions.

4.3.1 Double shell lining (DSL)

In the early days of sprayed concrete linings, the primary lining was regarded as temporary due to concerns over the long-term effects of some of the additives and the overall quality of the lining (e.g. encasement of steel mesh and arches in the concrete). A secondary lining was installed inside the primary lining to carry all permanent loads. This approach is still used in many cases and is called a "two-pass lining" or a double shell lining (DSL).

Secondary linings are normally formed of cast *in-situ* concrete, although sprayed concrete can be used, especially where the cost of formwork is high (e.g. at junctions or tunnels of varying shape). Since the secondary lining is placed inside a waterproof sheet membrane, it is typically designed to carry the water loading plus most or all of the ground loads (see sheet membranes above).

Cast concrete linings are placed within formwork (see Figure 4.10), and normal concrete technology is applied. The type of formwork depends on the specific requirements of each tunnel. Mobile steel formworks are used for longer tunnels with a constant cross-section. Timber formwork is cheaper in terms of materials but more labour intensive. Hence, it is used in countries where wage costs are low or for special cases, such as junctions, where it is not cost-effective to buy steel formwork.

Ideally, inner linings are designed to be unreinforced concrete. The shape of the tunnel can be chosen to minimise bending moments and, depending

Figure 4.10 Formwork for cast concrete secondary lining.

on the compressive hoop load in the tunnel, they may be low enough to be safely within the capacity of plain concrete. The risks of cracking due to thermal or shrinkage effects can be reduced by good mix design (e.g. using cement replacements like pulverised fly ash (PFA) to slow the hydration process) and casting the lining in short lengths (e.g. less than 10 m long bays). If there is no waterproofing membrane, it is sometimes advisable to install a plastic separation sheet to reduce friction on the contact with the primary lining. Again, this reduces the risk of cracking. Where lining loads are high, reinforcement is added to the secondary lining.

"Grey rock"

An alternative to ignoring the primary lining completely in the long-term design is to recognise that it remains in place and to ascribe the mechanical properties of a cohesionless gravel to it in the design calculations. This is the so-called "grey rock" design philosophy. In practice, this approach is only used in numerical analyses as it is too complex to incorporate into analytical solutions. The benefit in terms of the reduction in axial loads on the secondary lining is often very small but it may help because high "predicted" bending moments in the primary lining may disappear when it is turned into gravel in the analysis. Given the limited benefit, engineers may prefer to spend their energies demonstrating that the primary lining sprayed concrete is durable in the long term rather than juggling with "grey rock" parameters in numerical analyses. Table 4.7 contains some examples of these parameters.

4.3.2 Composite shell lining (CSL)

As noted earlier, an attraction of sprayed applied waterproofing membranes is the potential for load sharing between the primary and secondary linings – in a so-called composite shell lining – to form thinner and more economic linings. Whether or not these benefits can be realised depends heavily on the design approach. Early projects assumed zero bond in both shear and tension at the membrane (e.g. the Crossrail project in London, UK). The

Table 4.7 Design approaches for "grey rock" – degraded sprayed concrete

Approach	Values
50% degradation*	$E = 10$ GPa, $c = 2$ MPa and $\Phi = 45°$
Full degradation*	$E = 0.1$ GPa and zero strength
Channel Tunnel Rail Link	$c = 0$ MPa and $\Phi = 30°$; stiffness not stated
Terminal 5 project, UK	Ignore outer 75 mm of lining thickness but the remainder retains full strength; $E = 15$ GPa for long term in line with Eurocode 2

* presented by M. John at Nove Trendy v Navhovani Tunelu II seminar, Prague 2006.

natural consequence of this – coupled with an assumption that the water pressure acts on the membrane – is that a relatively thick secondary lining is required. One study of a tunnel in London inferred evidence of the long-term water saturation of primary lining from strain gauge data (over a period of more than 18 years), along with evidence of good durability of the primary lining from the stable stresses in the primary lining (Jones 2018).

Under this assumption, typically the secondary lining will be the same thickness in the traditional Double Shell Lining (DSL) approach (Bloodworth & Su 2018). However, in the case of Crossrail, there was substantially less reinforcement (i.e. ~ two-thirds less) in the secondary linings than there would have been if the DSL approach was used. Overall, because this variant of the CSL does not offer great savings, it is not recommended.

Composite lining action in a CSL

By now it is generally accepted that there is sufficient test data for some SAWMs to include meaningful values of bond at the interface of the membrane. Given that the bond and the membrane itself has finite properties which are lower than the concrete linings, the composite action is less than the ideal case. Numerous studies have examined the extent of composite action and the influences on it (Thomas & Pickett 2012, Su & Bloodworth 2014, Su & Bloodworth 2016, Su & Uhrin 2016, Jung et al. 2017a & b, Vogel et al. 2017, Bloodworth & Su 2018, Su & Bloodworth 2018). For the cases of shallow tunnels in soft ground, it appears that the strength of the bond – in both shear or tension – is well above the predicted stresses. The behaviour is dominated by the bond stiffness and particularly by the shear stiffness. Broadly speaking, the studies share similar conclusions on the extent of composite action. As an example, the short-term degree of composite action (DCA) has been found to vary from 0.50 to 0.85, considering the typical ranges of shear and normal stiffnesses of the membrane (Su & Bloodworth 2018). A full composite has a DCA of 1.0 while DCA = 0.0 for no composite action. Bloodworth and Su (2018) recommend halving the properties of the bond when considering long-term cases. This implies a relatively modest load sharing between the primary and secondary linings in the long term, i.e. more load is carried by the secondary. However, each case must be examined individually on its own merits by the design team.

In the case of a CSL, there are various ways to model the linings and the membrane in a numerical model. The most comprehensive method is to simulate the lining as zones with the membrane as zones (Jung et al. 2017a) or preferably an interface (Thomas & Pickett 2012, Bloodworth & Su 2018). Su and Bloodworth (2018) describe the procedure for this sort of numerical modelling, in the context of soft ground tunnelling.

The most influential parameter for determining load sharing is the shear stiffness of the interface. Bearing in mind that long term, partially saturated values for shear stiffness tend to be at the lower end of the range (i.e. around 0.2 GPa), based on the work of Su and others, one would expect a relatively weak composite action. However, this also depends on the other boundary conditions, and designers are advised to perform independent calculations.

As an alternative to modelling the two linings explicitly, some authors have explored the use of structural theory for laminates and attempted to derive the equivalent parameters for the monolithic lining (Jaeger 2016, Jung et al. 2017a & b, Thring et al. 2018). This is appealing as it opens the way for much simpler analyses. However, the equivalent thickness of the lining differs depending on whether one is matching deflections or stresses (Thring et al. 2018). The studies so far have not been validated for the stresses in the linings, and therefore this approach is not recommended.

4.3.3 Partial composite lining (PCL)

Thomas and Dimmock (2018) challenged the requirement for a structural secondary lining inside the SAWM. They proposed that – in structural terms – the primary lining would act in concert with the bonded membrane. The secondary lining inside then becomes a non-structural element with a range of possible functions such as fire protection for the membrane, aesthetics and carrying small fixing loads. Provided that there is a sufficient factor of safety against debonding, the membrane will transfer the water pressure acting on it back to the primary lining. In a study of lining thicknesses for a soft ground tunnel, Su and Bloodworth (2016) noted the desirability from a structural point of view of reducing the thickness of the secondary lining as far as possible.

This approach is also consistent with a current design approach for hard rock, drained tunnels. Holter (2015a) reported on three pilot studies from Norwegian tunnels, where a SAWM was applied in sections of tunnels. The thickness of the secondary lining inside the membrane ranged from 30 to 80 mm with a primary lining ranging from 180 to 330 mm. All of these tunnels are drained – i.e. the lining is not intended to be fully watertight, and water pressure is relieved by drains in the tunnel invert. Either the ground was relatively impermeable, or systematic pre-grouting was used to reduce the permeability of the rock mass to an acceptable level (see NFF Publication 20 (2011) for more information on systematic pre-grouting for rock tunnels).

4.3.4 Single shell lining – one pass lining

In its simplest form, the permanent lining can be formed using the so-called "single shell" or "monocoque" approach without any waterproofing membrane (Golser & Kienberger 1997). In fact, the "single shell" may consist of

Table 4.8 Examples of single shell SCL tunnels

Project	Type of tunnel	Reference
Munich sewer	Sewer	Gebauer 1990
Munich metro	Metro	Kusterle 1992
Heathrow Baggage Transfer tunnel	Non-public	Grose & Eddie 1996
Heathrow Terminal 5 "Lasershell" method	Water, road and rail tunnels	Hilar et al. 2005
SLAC Project	Research facility	Chen & Vincent 2011

several layers of sprayed concrete, placed at different times. However, the underlying principle is that all the sprayed concrete carries load over the life of the tunnel, and the different layers normally act together as a composite structure. This approach is common in certain sectors – notably on hydro-electric power projects – and in certain ground conditions – such as dry hard rock. Table 4.8 contains a short list of some examples.

Considering watertightness, Figure 4.11 shows that, if one considers a simple calculation of water inflow into a tunnel under various conditions (see Equation 4.1 after Celestino (2005) and Franzen 2005), currently single shell linings can only be used in relatively impermeable ground if the tunnel is to exceed ITA Class 3 (the minimum criterion for a railway running tunnel). The mass permeability of the ground would have to be less than 1×10^{-10} m/s. As an example, single shell linings have been

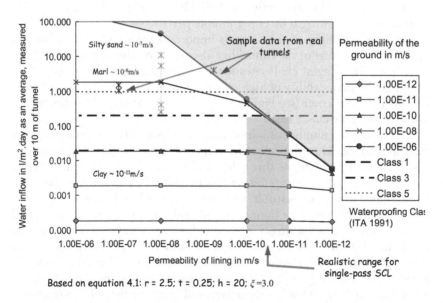

Figure 4.11 Water inflows into tunnels of various lining and ground permeabilities.

used successfully in London clay which has a permeability in the range of 5×10^{-11} to 5×10^{-12} m/s.

$$Q = \frac{2\pi k_l (h - r)}{\left[\ln\left(\frac{(r+t)}{t} \right) \right] + \left(\xi \frac{k_l}{k_g} \right)} \tag{4.1}$$

where Q is inflow in m^3/s per m of tunnel, k_l and k_g are the permeabilities of the lining and ground, h is the depth below groundwater level, t is the thickness of the lining and ξ is the "skin factor".

Figure 4.11 assumes that single shell linings (as a whole) can only achieve an impermeability that is about one order of magnitude less than the permeability of individual samples of sprayed concrete, i.e. 1×10^{-10} m/s to 1×10^{-11} m/s is achievable (see Figure 2.25 and Celestino 2005). This could be improved in the future – for example, with polymer additives (Bonin 2012). That said, there is also the option of improving the permeability of the ground too (e.g. by systematic grouting (Franzen 2005)).

It is also worth noting at this point that there is some debate about the acceptable levels of water ingress. The limits proposed by ITA (ITA 1991) and shown on Figure 4.11 may be too tight. Franzen (2005) noted that a single dripping spot per m^2 in a tunnel could result in about 1.5 l/m^2 per day (3 litres per minute per 100 m) while Celestino (2005) quoted 1 l/m^2 per day as an acceptable value for metro tunnels and 14 l/m^2 per day as the accepted limit in Norwegian subsea road tunnels (approximately 1.25 and 30 litres per minute per 100 m respectively). NFF (2011) suggested that acceptable limits range from 5 to 20 litres per minute per 100 m.

The impermeability of the lining is important with respect to achieving the specified dryness of the tunnel but also in preventing corrosion of reinforcement. The biggest challenges for a single shell lining lie in the joints. Although the permeability of high-quality sprayed concrete can be as low as 10^{-12} m/s or even lower (see Figure 2.25), the permeability of the lining as a whole is probably closer to 10^{-10} m/s due to inflow at joints.

Minimising the number of joints reduces the potential for inflow (e.g. by using a full-face excavation sequence rather than top heading and invert – if possible), but the scope for this is limited. Because of the large number of joints in the initial support (advance lengths during excavation are between 1 and 4 m), it is more sensible to ensure a good bond between adjacent panels of sprayed concrete than to install seals. The integrity of individual joints depends on achieving a good bond. Good spraying techniques are sufficient to ensure this. Sloping joints are sometimes preferred for ease of cleaning and for better compaction during spraying. They also have the merit of creating long and tortuous water paths. Staggering joints between the initial support and subsequent layers of concrete helps by extends the water path for incoming groundwater.

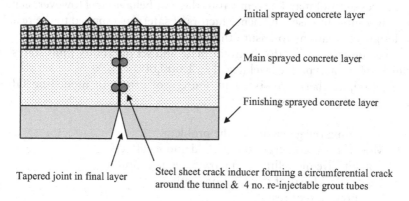

Initial sprayed concrete layer

Main sprayed concrete layer

Finishing sprayed concrete layer

Tapered joint in final layer

Steel sheet crack inducer forming a circumferential crack around the tunnel & 4 no. re-injectable grout tubes

Figure 4.12 Joint detail in sprayed concrete lining.

Suitable mix designs are discussed earlier in Section 2.1. Microsilica is often added because it fills pores and improves the density of the concrete.

At major joints – e.g. tunnel junctions – it may be necessary to install additional protection. Given their complicated shapes it is almost impossible to spray around conventional water-stops and encase them in concrete. Therefore, injectable grout tubes – see Figure 4.12 – are preferred for sealing major construction joints where differential movement may occur.

In terms of the design calculations for a single shell, the main difference arises from the composite action of the various layers of sprayed concrete. Few case studies exist to illustrate this aspect of the design. However, it appears that a modest bond strength (~0.5 MPa) is required in the radial direction between layers to permit composite action (Kupfer & Kupfer 1990). This is well within the achievable bond strengths for sprayed concrete (see Section 2.2.3). Having checked the integrity of the bond between layers, the full thickness of the sprayed concrete lining can then be used to carry the loads on it.

4.4 DESIGN TOOLS

A range of design tools exists for designers as outlined in the following sections. A more comprehensive discussion of the strengths, weaknesses and ranges of application of empirical, analytical, numerical and physical modelling tools for tunnel design can be found in the BTS Lining Design Guide (BTS 2004).

It is important to understand that all design tools are approximations of reality. The design tool simplifies the real case, analyses the simplified case and produces an answer – which is an estimation of how the real tunnel will behave. In very simple or well-understood cases, the estimation produced by the design tool may be identical to the real behaviour. However, generally this is not the case, and hence factors of safety are applied to the results of design calculations to ensure that the risk of tunnel failing to meet the specified design criteria (for both ultimate and serviceability limit states) is reduced to an acceptable level (e.g. 1 in 1,000,000).

In principle, there are six main sources of errors in modelling (after Woods & Clayton 1993):

- Modelling the geometry of the problem
- Modelling of construction method and its effects
- Constitutive modelling and parameter selection
- Theoretical basis of the solution method
- Interpretation of results
- Human error

and these should be borne in mind when choosing the design tools for a particular tunnel.

4.4.1 Empirical tools

Commonly used empirical design methods like the rock mass rating (RMR) system (Bieniawski 1984) and the Q-system (Barton et al. 1975, Grimstad & Barton 1993) have been developed for blocky or hard rock tunnels. They are quick and easy to use, at least for those with experience in estimating the input parameters. These methods employ a combination of parameters such as the strength of the rock, its quality (using rock quality designation (RQD) values), joints (number of sets, frequency, spacing and condition) and groundwater conditions to produce a rock mass classification. From design charts or tables, the support measures required are quantified, based on the product of these parameters.

However, there are limitations to these tools. For example, they are based on drill and blast tunnels. In TBM-driven tunnels the rock is less disturbed by excavation and so requires less support than might be predicted. On the other hand, there may be particular requirements for a project that mean more support is required than is suggested. Furthermore, these tools do not give an indication of the factor of safety related to the proposed support. These methods, and others, are reviewed in Hoek and Brown (1980) and Hoek et al. (1998).

Empirical methods for soft ground tunnels are rarely used now.

4.4.2 Analytical tools

These include continuum "closed-form solution" models (e.g. Curtis & Muir Wood) and Panet's convergence confinement method (CCM) – see also Table 4.2. The continuum analytical methods are relatively simple and provide information on stresses in the lining and its deformation. These are often used in the early stages of design. Some of them may be extended to include features such as plasticity in the ground or the timing of placement of the lining.

However, they share several fundamental limitations: they assume plane strain or axisymmetry, and the solutions are almost invariably developed only for circular tunnels, constructed in full-face excavation in homogeneous ground. This is a major weakness from the point of view of designing SCL tunnels, given the non-circular shapes and excavation sequences used. Furthermore, the modelling of soil–structure interaction is limited yet this is fundamental to all tunnels. Many form solutions in their basic forms make no allowance for stress redistribution ahead of the face, but sometimes this can be incorporated using a relaxation factor – see Table 4.1.

For the case of blocky rock, limit equilibrium calculations can be used to calculate the bolting required to support individual wedges. Typical factors of safety are 1.5 for grouted and 2.0 for ungrouted rock bolts (Hoek & Brown 1980). However, in practice these are performed by computer rather than by hand (e.g. using programmes such as UNWEDGE). Similar calculations for a variety of failure mechanisms can be performed to determine the required thickness of sprayed concrete between rock bolts in rock support (Barrett & McCreath 1995). The governing failure mechanism has been found to be generally a loss of adhesion to the rock, followed by flexural failure of the sprayed concrete (Barrett & McCreath 1995). If debonding does not occur, then shear failure is the most likely failure mechanism (Sjolander et al. 2018).

4.4.3 Numerical modelling

To overcome the limitations of empirical and analytical tools, one must turn to numerical methods, such as the finite element (FE) and finite difference (FD) methods, which can model the full complexity of an SCL tunnel explicitly. Ideally, the design calculation will consider explicitly the following:

- The construction sequence and the three-dimensional stress redistribution around the tunnel face
- The age- and time-dependent behaviour of the sprayed concrete
- Nonlinear material behaviour of the sprayed concrete and the ground[2]

Despite their advantages, numerical models are still merely approximations of reality and it is important to understand the limitations of the design models. Sound engineering judgement remains the key to good design. The more simple design tools still have a vital role to play as a check on the results from numerical models. Where possible the models should also be calibrated with data from site. Further advice on numerical modelling can be found in the BTS guide (2004). Some of the issues related to modelling composite linings have been discussed in Section 4.3.2. In the sections below comments have been made on some of the main sources of errors in numerical modelling and how to avoid them.

Modelling the geometry of the problem

The most obvious question in the numerical modelling of tunnels is whether to use 2D or 3D analyses. The value of using 3D models and avoiding such correction factors is increasingly recognised (Haugeneder et al. 1990, Hafez 1995, Yin 1996, Burd et al. 2000, Thomas 2003, Jones 2007), not least because the situation becomes considerably more difficult as more complex models are analysed (such as cross-sections with multiple headings or coupled consolidations analyses (e.g. Abu-Krisha 1998)). While 3D models are still too time-consuming for general use, they are being used on projects to determine input parameters (i.e. the relaxation factor) for 2D analyses or for complex cases such as junctions.

As an aside, a longitudinal plane strain model of a tunnel is not valid because this represents an infinitely wide slot, cut through the ground. Similarly, for shallow tunnels (i.e. C/D ~ 2), axisymmetry is not a valid assumption (Rowe & Lee 1992).

Where there are adjacent structures (both above and below ground) or features such as slopes, it may be important to model these too.

Modelling of construction method and its effects

SCL tunnels are built in a sequential process that often involves a subdivision of the tunnel face. This should be modelled so that the stability of these intermediate stages can be checked. Failure usually occurs at one of these stages rather than after the full tunnel lining has been installed. A more detailed discussion can be found in Section 5.8. All elements of the support system should be modelled – either explicitly or implicitly. For example, spiling or forepoling cannot be modelled in a 2D model explicitly, but the effect can be incorporated by enhancing the properties of the relevant area around the tunnel. Other construction activities may also be relevant and therefore should be modelled. One example of this is compensation grouting which can impose additional loads on the lining – see Section 6.8.6.

Constitutive modelling and parameter selection

Typically, relatively simple models are used for the sprayed concrete lining. The norm is to assume a homogeneous, isotropic, linear elastic constitutive model, albeit including some variation in elastic modulus with age. It is usually assumed that the lining has been constructed to the exact (nominal) geometry specified, even though this method is known to be vulnerable to poor workmanship. In contrast, segmental linings consider construction defects as a matter of course in the design, even though they are arguably less prone to poor workmanship.

The choice of constitutive model for the sprayed concrete has been found to affect the predicted stress distribution in the lining (Thomas 2003). This seems to be more pronounced if the utilisation factor in the lining exceeds 50%. Assuming that the tunnel face is stable, the loads in the lining and its movements are governed by the relative stiffness of the ground and lining, since this is a soil–structure interaction problem. Bending moments have been found to be more strongly influenced than hoop forces by the constitutive model (both for the sprayed concrete and ground). A discussion of possible constitutive models for sprayed concrete is contained in Chapter 5.

Looking more broadly, the constitutive model for the sprayed concrete may not have a large influence on the far-field behaviour of the ground. As might be expected, the assumed *in-situ* stress state and the constitutive model for the ground both can have a considerable influence on the predicted loads on the lining (Thomas 2003, Jones 2018). There is a growing appreciation of the need to model the ground with sophisticated constitutive models.

Theoretical basis of the solution method

In Chapter 3, the concepts of continua – soft ground or hard rock – and discontinua – blocky rock – have already been introduced. The numerical modelling programme should be appropriate for the type of ground under consideration. FE and FD programmes model continua while discrete element method programmes (e.g. UDEC and 3DEC) or UNWEDGE model discontinua. Bedded beam models are used less and less these days, because of their limited ability to model the soil–structure interaction. They have been largely superseded by 2D numerical models.

There are ways to model discontinua using conventional FE or FD programmes. For example, one can use the Hoek–Brown failure criteria to approximate the behaviour a jointed rock mass in a continuum model or one can introduce interface elements to model major discontinuities.

4.4.4 Physical tools

Trial tunnels are used occasionally for research or when SCL tunnelling is proposed for a new or particularly difficult area. Trial tunnels can provide the most readily accessible and realistic data on the performance of tunnel linings, although at considerable expense. Examples from the UK include: the Kielder experimental tunnel (Ward et al. 1983), Castle Hill trial heading (Penny et al. 1991), the trial tunnels for Heathrow Express (Deane & Bassett 1995) and Jubilee Line Extension (Kimmance & Allen 1996). On a broader note, the results from monitoring during and after construction bolster the general understanding of tunnel behaviour (e.g. the loads on linings (Jones 2018) or the use of data from the HEX and JLE projects to underpin the design of SCL tunnels for Crossrail in London (Goit et al. 2011)) and can be used to enhance empirical design methods.

Small scale and full-scale models are rarely used directly in the design of sprayed concrete linings but have been used in research. Large scale models of tunnel linings have been constructed and tested to examine behaviour under working and collapse loads (e.g. as part of a recent Brite Euram project[3] (Norris & Powell 1999)). Other examples include: Stelzer & Golser (2002); Stark (2002) and Trottier et al. (2002).

4.5 CODE COMPLIANCE

All tunnels must be designed in accordance with the relevant national design standards. However, most countries do not have specific design codes for underground excavations. Hence, code compliance becomes a "grey area". Many design codes are based on the principle of ultimate limit state (and serviceability limit state) – e.g. in Europe and North America. The principle states that the probability of the loads exceeding the strength of the structure should be so small as to be negligible. The "worst credible" loads are multiplied by a load factor (typically 1.35 design according to Eurocodes). The specified strength of the structure is divided by a material factor (typically 1.50 for concrete in design according to Eurocodes). The factored strengths should always be higher than the factored loads. The combined load and material factors provide an overall factor of safety that is typically greater than 2.0. Furthermore, the structure should have sufficient redundancy that it does not fail in a sudden, brittle manner.

Compliance with the basic principles seems straightforward. Design codes exist for reinforced concrete structures (e.g. Eurocode 2 (2004) and ACI 318) and these guide calculations to check compliance under different forms of loading (e.g. axial force, bending, shear, etc.). A tunnel lining in soft ground could be seen as a structural member under combined axial force and bending moments. Shear loading is more important in blocky rock tunnels. Existing design codes for concrete can be used directly for any aspects that are independent of the fact that the concrete has been sprayed or is a tunnel lining – for example, determining cover to reinforcement.

The first difficulty arises in the estimation of the credible ground loading. Section 4.2.1 describes two approaches for determining ground loads in design. To obtain the "worst credible" estimate of the loads, "moderately conservative" geotechnical parameters are first used to estimate the loads which are then multiplied by a load factor (typically 1.35). This is in line with Eurocode 7 Design Approach 2.

Secondly, design codes for reinforced concrete are written for conventionally placed concrete at ages greater than seven days. To comply with the normal factors of safety and the stress–strain behaviour set out in the codes, typically the utilisation factors[4] in the tunnel lining should be less than 39% ($0.8\ f_{cu} \div \{1.35\ (\gamma_f) \times 1.5\ (\gamma_m)\}$). While some codes (e.g. BS8110 Part 2 1985) permit some latitude on the basis of experimental data and engineering judgement, it would still be difficult to justify utilisation factors that exceed 55% ($1.0\ f_{cu} \div \{1.2\ (\gamma_f) \times 1.5\ (\gamma_m)\}$) for short-term loads near the face.[5] As Figure 4.13 shows the utilisation factor in the lining may be greater than 55% within the first few metres of the lining. The concrete here is less than seven days old .

This raises two questions. Should normal design codes be applied at these ages? How can one prove the safety of the tunnel in this critical area?

One could assume that the codes do not apply and resort to other means to prove the stability of the tunnel, e.g. empirical methods based on the stability number, N, or limit equilibrium solutions. The limitations of this approach and the difficulty in application to SCL tunnels have been discussed earlier. One could strengthen this approach by combining it with a process of risk management which culminates in the use of monitoring data during construction to verify that the tunnel is behaving as intended in the design (Powell et al. 1997), although one does not have the opportunity to take many readings in the critical area which is near the tunnel face. The

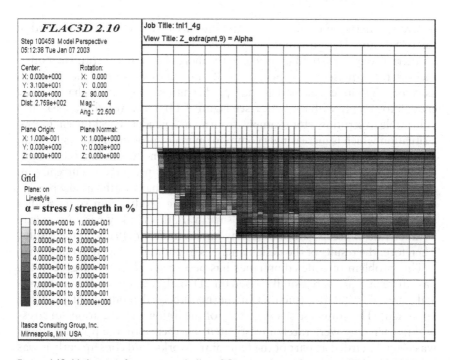

Figure 4.13 Utilisation factors in a shallow SCL tunnel in soft ground (Thomas 2003).

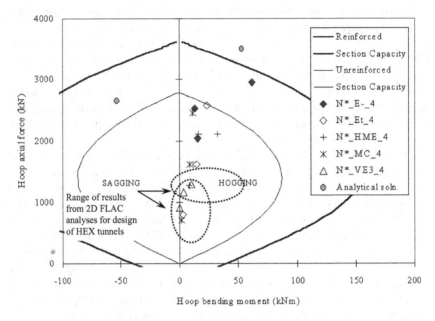

Figure 4.14 Results from numerical model of a shallow SCL tunnel in soft ground (Thomas 2003).

differences between this approach and the classical NATM are discussed in the BTS Lining Design Guide (BTS 2004).

Alternatively, one could use some of the more elaborate constitutive models such as those presented in Section 5. By taking advantage of factors such as creep, the numerical models may predict utilisation factors that are low enough to comply with the codes (see Figure 4.14). However, this approach is vulnerable to scrutiny – for example, from an independent checking engineer – a situation which is increasingly common on major projects. Few projects can afford an extensive pre-construction test programme to validate the parameters for these constitutive models. At the same time, there is currently an insufficient pool of data from which to determine many of the model parameters with the certainty normally required. The results from some studies (e.g. Thomas 2003) have shown that small variations in key parameters (e.g. advance rate or creep parameters) may have significant effects on the results. Therefore, the results of the design analyses might remain open to question.

This problem of code compliance has been concealed to a degree in the past since designs were usually based on 2D analytical methods with empirical correction factors (such as in the hypothetical modulus of elasticity approach). The complete stress history of the lining and variations in stress within each ring were rarely investigated. Furthermore, the primary lining was often regarded as part of the temporary works, and consequently it was subjected to less rigorous scrutiny.

A pragmatic approach may be:

- To acknowledge that, while the principles of conventional design codes apply to the heading of an SCL tunnel, the detail may not apply at very early ages
- To embrace the new information provided by 3D numerical analyses as a complement to existing design approaches
- To accept that numerical models are still only an approximation of reality – so sound engineering judgement remains as important as ever
- And to validate the performance of the tunnels by monitoring during construction

4.6 CONTINUITY BETWEEN DESIGN AND CONSTRUCTION

Surprising as it may seem, a bundle of drawings and a hefty sheaf of contract documents are not enough to convey everything that the designer has in mind for a tunnel. A certain amount of the design philosophy – at least in current practice – is not captured and passed on to the construction team. This creates a risk that changes during construction may compromise the design criteria or that abnormal behaviour of the structure will not be identified before lasting damage is done.

Several simple steps can be taken to mitigate this risk. Firstly, a design report can be included with the construction drawings. This brief report describes the design criteria, the philosophy adopted to meet them and the expected performance limits. The report may include details of the design methods.

A good dialogue between construction teams and designers improves the mutual understanding and leads to more economic designs and greater safety during construction. The benefits of the presence of a representative of the designer on site are discussed further in Section 7.4.

NOTES

1. The term "active" design has been used to describe the same process in rock tunnelling (Grov 2011).
2. NB: soft ground often exhibits complex material behaviour, with features such as nonlinear stress–strain behaviour, plasticity, variable K_0 values, anisotropy and consolidation (e.g. London Clay – see Thomas 2003).
3. BRITE EURAM BRE-CT92-0231 project on New Materials, Design and Construction Techniques for Underground Structures in Soft Rock and Clay Media, part funded by the Commission of the European Communities, 1994–98.

4. NB: the calculated utilisation factors from the numerical model are based on deviatoric stresses whereas BS8110 Part 1 (1997) considers stress components in each direction separately – e.g. the hoop axial load vs. f_{cu}. Eurocode 2 (2004) (Cl. 3.1.9) permits the increase in strength due to confinement to be considered.

5. A partial factor of safety of 1.20 is commonly used for temporary loads.

Chapter 5

Modelling sprayed concrete

As Fourier said, *"nature is oblivious to the difficulties that it poses for mathematicians"*. Similarly, sprayed concrete is blissfully unaware of the headaches it causes tunnel engineers. As explained in Section 2.2, sprayed concrete exhibits complex behaviour:

- The mechanical properties of sprayed concrete change considerably as the hydration reaction proceeds and a hardened concrete is formed (i.e. it *ages*).
- The rate of change of properties is rapid initially, due to the accelerated hydration and slows with age.
- In its hardened (and hardening) form, sprayed concrete is a nonlinear elastoplastic material when under compression.
- In tension it is initially a linear elastic material which then fails in a brittle manner.
- To overcome the brittle failure, tensile reinforcement is often added.
- Sprayed concrete exhibits shrinkage.
- The creep behaviour of sprayed concrete can be pronounced and changes with age.
- Different mixes can have substantially different mechanical properties.

These features must be borne in mind during the design process but not all of them will necessarily be relevant in every case, and so it may be possible to simplify the design. Such simplifications should be made on a reasoned basis and not just to make the designer's life easier. Otherwise the design may become unsafe or over-conservative. In contrast to the geotechnical model, relatively crude models are normally used for sprayed concrete (e.g. Bolton et al. 1996, Watson et al. 1999, Sharma et al. 2000). Therefore, it is perhaps unsurprising that there is often a significant discrepancy between the behaviour predicted by numerical analyses and that observed on site (Addenbrooke 1996, van der Berg 1999). The results of design analyses have been shown to depend strongly on how the sprayed concrete is modelled, although the difference is most pronounced when the lining is heavily loaded – i.e. the stress is higher than 50% the strength (Thomas 2003).[1]

Table 5.1 Common design parameters for sprayed concrete

Parameter	Description	Typical range	Type of ground
E	Elastic modulus	30 to 35 GPa at 28 days NB: E varies widely, from zero at age = 0	Soft/blocky
f_{cyl} (or f_{cu})	Compressive strength (from cylinders or cubes)	25 to 40 MPa at 28 days NB: f_{cu} varies widely, from zero at age = 0	All
f_{cu} (t)	Compressive strength (from Hilti gun tests)	J2 – ÖBV 1998	All
v	Poisson's ratio	0.20 *	Soft/blocky
–	Bond strength	0.125 to 0.35 MPa after spraying, rising to 0.5 to 1.4 MPa at 28 days	Blocky/hard
f_{Rlk}	Flexural strength – for steel fibre sprayed concrete	1.50 to 2.50 MPa at 28 days	Blocky/hard

* This is a reasonable value except when the concrete is close to failure (Chen 1982).

Improving the modelling of the sprayed concrete could make the prediction of the behaviour of these tunnel linings more reliable.

Table 5.1 lists some of the design parameters for sprayed concrete that may be required as well as the type of ground where they may be relevant. Section 2.2 provides more detail on each parameter.

There is no single perfect constitutive model for sprayed concrete. This chapter will present some of the models that exist and highlight their strengths and limitations. The opportunities to explicitly incorporate the more complex aspects of this behaviour into design calculations are very limited in analytical design tools and non-existent in empirical tools. Hence the comments below are mainly angled towards the numerical modelling of sprayed concrete structures.

5.1 LINEAR ELASTIC MODELS

The most commonly used model is a linear elastic one with a constant stiffness, because of its simplicity and computational efficiency. Typically, elastic models predict axial forces and bending moments in linings that are unrealistically high compared with field data from strain gauges and pressure cells (Golser et al. 1989, Pöttler 1990, Yin 1996, Rokahr & Zachow 1997). This is no surprise since sprayed concrete only behaves in a linear elastic manner up to about 30% of its uniaxial compressive strength (Feenstra & de Borst 1993, Hafez 1995), and the stiffness varies considerably during the early age of the sprayed concrete.

A logical improvement on a simple elastic analysis is to incorporate the increase in magnitude of the stiffness with age (see Figure 2.14 and

Appendix A). In most cases, at the design stage, there is no experimental data for the stiffness (e.g. elastic modulus, E, and Poisson's ratio, v) at different ages of the sprayed concrete mix in question. Instead the elastic modulus may be estimated from the strength of the sprayed concrete (Chang & Stille 1993; see also Appendix A), using the equation:

$$E = 3.86\sigma^{0.60}$$
(5.1)

where σ is the uniaxial compressive strength. If the elastic modulus is known at 28 days (E_{28}), the value at other ages may be estimated from any one of a number of equations (see Appendix A), for example:

$$E = E_{28} \cdot \left(1 - e^{-0.42t}\right)$$
(5.2)

where t is the age in days and E_{28} is the stiffness at 28 days (Aydan et al. (1992b). The Poisson's ratio can be assumed to be constant with age and equal to 0.20. In numerical models, this is often implemented as a "stepped" approximation of the curve, since the excavation sequence is modelled as a series of steps. The increase in stiffness with age will lead to irrecoverable strains on unloading at later times (Meschke 1996).

Influence on the predictions of numerical models

There is plenty of evidence to support for the use of an age-dependent linear elastic model for sprayed concrete. In a 3D numerical model, Berwanger (1986) also found that the ultimate stiffness of the sprayed concrete had a limited influence on surface settlement but it did have a large influence on the stresses in certain parts of the lining, notably the footing of the top heading. Similarly, in 2D numerical models, Pöttler (1990), Huang (1991), Hirschbock (1997) and Cosciotti et al. (2001) found that increasing the stiffness of the lining increased the stresses in it. Considering a tunnel constructed in one stage, in a 2D analysis, Hellmich et al. (1999c) stated that the stiffness of the lining affected only the hoop bending moments and not the hoop forces. However, later in the same paper they show that the axial forces are lower when rate of hydration is slower. By extrapolating their finding that the stresses in the lining are influenced by how long it takes for the lining to start to carry meaningful loads, it would seem that in a multi-stage construction sequence the rate of growth of stiffness will be an important factor.[2]

Several authors have suggested that the relative stiffness of the lining compared with that of the ground would affect the amount of influence of a time-varying modulus (Hellmich et al. 1999c, Cosciotti et al. 2001). In their set of 2D analyses, Hellmich et al. (1999c) found that the rate of stiffening of the lining was important in ground that creeps at a similar rate to the sprayed concrete, but when the creep in the ground is much slower, the rate of hydration of the concrete had virtually no effect on the axial forces and moments. In a detailed study using 3D numerical models, Soliman et al. (1994) reported that a variable elastic modulus led to significantly larger lining deformations (20–30% more) and lower bending moments (reduced by up to 50%) compared to a constant elastic modulus. The thrust loads were slightly lower – reduced by about 20% – and hence the stresses in the ground were not increased by much. Similarly, Jones (2007) found that in a 3D model of shaft-tunnel junction, the bending moments were affected more than axial forces when varying the ground constitutive model and lining thickness – i.e. the relative stiffness of lining and ground. This may well explain why surface settlements seem to be independent of the constitutive model of the lining (see also Moussa 1993).

In conclusion, in a multi-stage construction sequence, using an age-dependent elastic modulus for the lining will result in lower stresses in the lining (e.g. see Figure 5.1). The bending moments are reduced more than the hoop forces (e.g. see Figure 5.2). The reduction appears to be due more to the lower stiffnesses during early loading (i.e. ages less than 48 hours), compared to a (high) constant stiffness model, than how the stiffness develops beyond that period.[3] If the lining is not heavily loaded (i.e. loaded to a utilisation of 40% or less), it is probably not necessary to use a more complicated model than the age-dependent linear elastic model. Sprayed concrete behaves linearly up to stresses of 30 to 40% of the uniaxial compressive strength (see Section 2.2.4).

State of the art SCL designs often employ a relatively crude usage of the hypothetical modulus of elasticity (HME). It is suggested that there are

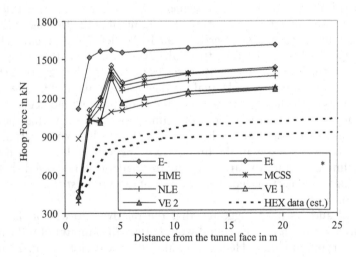

Figure 5.1 Hoop axial force in crown vs. distance from face for different sprayed concrete models (Thomas 2003).

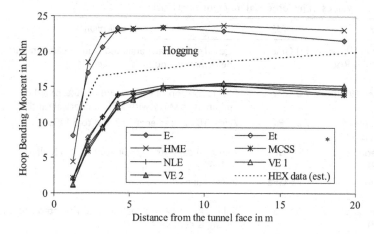

Figure 5.2 Hoop bending moment in crown vs. distance from face for different sprayed concrete models (Thomas 2003).

sufficient data to use more realistic constitutive models – e.g. an ageing linear elastic perfectly plastic model based on site data for strength development with age (and stiffness estimated from the strength) or an ageing linear elastic model with the stiffness reduced to account for creep – a *refined* HME method. The latter was used very successfully on the Crossrail project in UK (Goit et al. 2011).

5.2 HYPOTHETICAL MODULUS OF ELASTICITY (HME)

A widely used and very successful attempt to improve numerical modelling came in the form of the hypothetical modulus of elasticity (HME) method (Pöttler 1985). In this, several reduced values of elastic modulus are used in an analysis. Typical values are shown in Table 5.2. The "softer" lining leads to more realistic results – larger lining deformations and lower stresses – without excessive computation time. Though the concept of an effective modulus is not new to creep analysis (e.g. BS8110 Part 2 1985),[4] the HME is intended to account for ageing of the elastic stiffness, shrinkage and 3D effects as well as creep. According to the original formulation:

$$E_{HME} = E_T \cdot f_v \cdot f_{s,k} \cdot f_{vv} \qquad (5.3)$$

where E_T = the stiffness at the time in question, f_v = correction factor for the age-dependent stiffness during the loading up to the time in question, $f_{s,k}$ = correction factor for creep and shrinkage and f_{vv} = the crown

Table 5.2 Values of hypothetical modulus of elasticity

Project	HME	Application
Channel Tunnel (Pöttler & Rock 1991)	1.0 GPa	Age < 14 days; back analysed from measurements of deformation and pressure cells
(Pöttler 1990)	7.0 GPa	Plus relaxation of the ground of 30%; based on parametric study with a 2D numerical model
CTRL North Downs	7.5 GPa	Age < 10 days; strength limited to 5 MPa
(Watson et al. 1999)	15.0 GPa	Age > 10 days; strength limited to 16.75 MPa (= 0.67 f_{cu})
Heathrow Express	0.75 GPa	Initial value
(Powell et al. 1997)	2.0 GPa	Value after adjacent section is constructed and until lining is complete
	25.0 GPa	Mature sprayed concrete
(John & Mattle 2003)	1.0–3.0 GPa	For sprayed concrete with a one-day strength < 10 MPa and light reinforcement (i.e. high creep potential)
	3.0–7.0 Gpa	For sprayed concrete with a one-day strength > 10 MPa and moderate to heavy reinforcement (i.e. low creep potential)
	15.0 GPa	Mature sprayed concrete
Crossrail (Goit et al. 2011)	varies	Stiffness determined for each excavation stage depending on the age at the end of the stage, based on Chang & Stille (1993), divided by 2.0 to allow for creep.

deformation occurring before lining placement as a fraction of total deformation of ground at the crown – i.e. the effects of 3D stress redistribution and timing of placement. Sometimes the HME includes an allowance for the nonlinear stress–strain behaviour of the concrete (e.g. John & Mattle 2003).

The correction factors in Equation 5.3 require knowledge of how the lining and ground will deform as well as how creep will alter the stresses in the lining. While the first can be estimated using various analytical tools, the second cannot. Hence the choice of the values of HME is usually empirical. Sometimes the HME is combined with relaxation of the ground stresses ahead of the face in a 2D numerical model (i.e. $f_{vv} = 1$ but the relaxation is catered for explicitly in the model – e.g. Pöttler 1990 or John & Mattle 2003). While John and Mattle (2003) provide the most detailed description of how to choose the values of the HME, their approach should be treated with caution as it seems to rely on knowing the answer before the calculation is done – i.e. one must know how heavily the lining will be reinforced, how fast the ground will apply the load to the lining and which is the most critical stage of the tunnel construction in order to select the correct HME value.

Influence on the predictions of numerical models

Clearly, given the low stiffnesses at early ages, one would expect that using the HME approach would result in significantly lower predictions of stresses than using a (high) constant stiffness model.

5.3 NONLINEAR STRESS–STRAIN BEHAVIOUR

As noted earlier, the stress–strain curve for concrete is nonlinear at stresses above about 30% of its uniaxial compressive strength (see Figure 2.5). This nonlinearity can be implemented in the theoretical frameworks of either strain-hardening plasticity or nonlinear elasticity.

5.3.1 Nonlinear elastic models

Models, such as the Cauchy, hyperelastic and hypoelastic models, attempt to replicate the nonlinear stress–strain behaviour of concrete. This nonlinear behaviour begins at relatively low stresses[5] and is due to microcracking at the interface between the aggregate and cement paste, which themselves are still responding elastically (Neville 1995). Since this plastic deformation lies behind the nonlinearity, plasticity models are required if unloading occurs. However, if unloading can be neglected, nonlinear elastic models represent an economic means of modelling the nonlinear response of concrete to loading and a significant improvement on linear elastic models (Chen 1982). Consequently, they have been widely used in the analysis of concrete structures, although they have been rarely used in the analysis of sprayed concrete tunnels.

Specific nonlinear elastic models used for the analysis of SCL tunnels include: Saenz's equation (see Chen 1982) which Kuwajima (1999) found fitted experimental data for stress–strain curves well; the rate of flow method (see Section 5.6.5) and the parabolic equation below (Moussa 1993).[6]

$$\sigma_c = f_c \cdot \left(\frac{\varepsilon_c}{\varepsilon_1} \right) \cdot \left(2 - \frac{\varepsilon_c}{\varepsilon_1} \right) \tag{5.4}$$

where f_c is the peak stress, ε_1 is the strain at peak stress and σ_c and ε_c are the equivalent uniaxial stress and strain respectively.

In these models, the behaviour of concrete is treated as an equivalent uniaxial stress–strain relationship. Biaxial effects may be accounted for in the tangent moduli.

The Kostovos–Newman model

This model (Kotsovos & Newman 1978, Brite Euram C2 1997) is formulated in terms of octahedral stresses, and so, unlike the others, it has the advantage of being designed for generalised states of stress. Although quite lengthy in its formulation, another advantage of this model is that all the parameters can be determined from the uniaxial compressive strength of a cylindrical sample and its initial stiffness – see Appendix B.

Considering it in more detail, its other key advantages are that:

- The model includes the effects of deviatoric stress on hydrostatic strains.
- The model has been shown to agree well with existing test data for concrete and sprayed concrete (Brite Euram C2 1997, Eberhardsteiner et al. 1987, Thomas 2003), and hence it has been recommended for use in modelling mature sprayed concrete (Brite Euram 1998).
- The increase in strength with increasing hydrostatic stress is accounted for, and the predicted failure surface agrees better than the Mohr–Coulomb model – see Figure 2.12.

However, it should be noted that, as with nonlinear elastic models, it is only valid up to the point of onset of ultimate failure, which is at about 85% of the ultimate strength.

The formulae for the tangent shear and bulk moduli in this model are shown below with full details of all parameters explained in Appendix B.

$$G_{\text{tan}} = \frac{G_0}{\left(1 + \left(C.d.\left(\frac{\tau_0}{f_{\text{cyl}}}\right)^{d-1}\right)\right)} \qquad (5.5)$$

$$K_{\text{tan}} = \frac{K_0}{\left(1 + \left(A.b.\left(\frac{\sigma_0}{f_{\text{cyl}}}\right)^{b-1}\right) - \left(k.l.m.e\left(\frac{\tau_0}{f_{\text{cyl}}}\right)^{n}\right)\right)} \qquad (5.6)$$

Since this is a tangent modulus model, its accuracy depends on the size of the load increments in comparison to the peak strength. Thomas (2003) suggested that a minimum of seven equal increments is required for acceptable performance. The size of load intervals in numerical models is a trade-off between accuracy and speed. Minor modifications have been made to the formulae to extend them to strengths less than 15 MPa. The shear modulus in the original formulation by Kotsovos and Newman does not decrease to zero at the peak strength. To overcome this, Thomas (2003) proposed that modulus is reduced gradually to 5% of G_0 (the initial shear modulus) as the actual shear stress approaches the peak shear stress (at 85% of the peak stress). Above the same point the bulk modulus is reset to 0.33 K_0 (a third of the initial bulk modulus) in line with recommendations from Gerstle (1981).

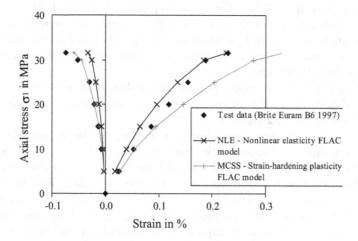

Figure 5.3 Back analysis of a uniaxial compression test on sprayed concrete (Thomas 2003). Input data for both models: f_{cyl} = 31.5 MPa, v = 0.2, E_0 = 18.80 GPa.

Figure 5.3 and Figure 5.4 suggest that this constitutive model functions well both under uniaxial and triaxial loading. Despite using the same input parameters as a strain-hardening plasticity model, Thomas (2003) found that the nonlinear elastic model agrees better with the test data in the triaxial case (see Figure 5.4). The nonlinear elastic model had been optimised to fit the test data by altering the point at which the moduli reduce to the low

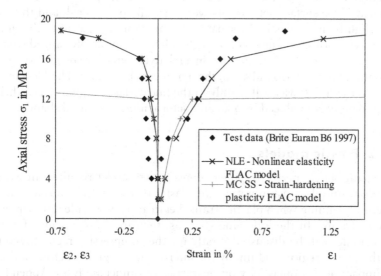

Figure 5.4 Back analysis of a triaxial compressive test on sprayed concrete (Thomas 2003). Input data for both models: f_{cyl} = 9.5 MPa, v = 0.2, E_0 = 8.45 Gpa.

values, as described above. This leads to predicted stresses that exceed the model's own estimate of the strength of the concrete (by about 12%) given the confining stresses and the uniaxial strength. This is possible because it is an elastic model, whereas the strain-hardening plasticity model is capped at its predicted peak strength.

Ageing is a major complication in the implementation of the constitutive models for sprayed concrete, but it can be successfully handled by the Kotsovos and Newman model.

The effects of loading, unloading or reloading can be included by using the Masing rules for loading cycles (Dasari 1996). Unloading is probably of limited relevance to the lining of a single tunnel constructed on its own with a top heading, bench and invert excavation sequence, since little unloading would be expected to occur. However, it is likely to be of relevance where other tunnels are constructed near to an existing tunnel, at junctions and in more complex excavation sequences. Since loading is determined on the basis of changes in octahedral deviatoric stress alone, changes in hydrostatic stress are not recognised in terms of loading/unloading. However, a tunnel lining is predominately in a biaxial stress state, and so the most likely changes in load are primarily deviatoric ones, rather than purely hydrostatic.

Influence on the predictions of numerical models

In a 2D analysis of a shallow tunnel in soft ground, Moussa (1993) found that incorporating nonlinearity into the elastic model resulted in a reduction in hoop forces of about 20% and a reduction of up to 50% for bending moments. The surface settlements were virtually unchanged, and there was only slightly more plastic deformation in the ground adjacent to the tunnel. The precise utilisation factors are unclear but they probably ranged from 40 to 80%. Thomas (2003) found that in a relatively lightly loaded lining (with a utilisation factor of around 40%) the nonlinear elastic model had less impact because the lining was still mainly in the linear elastic region. Nevertheless, hoop forces were reduced by 5%, and deformations increased slightly.

5.3.2 Plastic models

A general elastic perfectly plastic constitutive model requires an explicit stress–strain relationship within the elastic region, a yield surface (failure criterion) defining when plastic strains begin and a flow rule (governing the plastic strains). In the following sections, the three components of a plasticity model will be discussed, firstly for the compressive region, secondly for the tensile region and finally for intermediate regions, before outlining the impact of elastoplasticity on the results of tunnel analyses. Appendix C contains details of plasticity models already used for analysing sprayed concrete tunnel linings.

Elastic behaviour

Isotropic linear elasticity is generally assumed up to the yield point (about 30 to 40% of the compressive strength).

Yield criteria

Since the first one parameter yield criteria were proposed by Rankine, Tresca and von Mises, many others have been formulated (Owen & Hinton 1980). More and more complex criteria have been proposed to match experimental data more accurately over a wider range of stresses. In the case of concrete, the two parameter Mohr–Coulomb[7] and Drucker–Prager yield criteria have been used often in the past (see Figure 5.5). New yield criteria have been developed which can replicate the curved nature of the yield surface meridians (see Figure 2.12) and also the shape of the surface in the deviatoric stress plane, which is initially almost triangular but tends to an almost circular shape at high hydrostatic stresses (Hafez 1995, Chen 1982). Curved yield surfaces are also advantageous since the corners and edges are difficult to handle in numerical analysis (Hafez 1995). Hence, for the purposes of analysis of sprayed concrete tunnel linings, the Drucker–Prager criterion has been the most widely used (see Appendix C).

Considering a moderately heavily loaded tunnel, where the principal stresses in the lining might be 8, 3 and 0.5 MPa and the 28-day strength equals 25 MPa, the normalised octahedral stress, σ_{oct}/f_{cu}, is only 0.15 which is quite low. So, the assumption of straight meridians in the Drucker–Prager criterion is reasonable. The Drucker–Prager criterion can also be amended to reflect the increase in yield stress in biaxial states of stress (Hafez 1995, Meschke 1996).

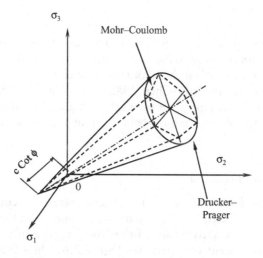

Figure 5.5 Yield surfaces in 3D stress space.

However, the shape of the Mohr–Coulomb criterion in the deviatoric plane agrees better with the almost triangular yield surface suggested by test data in the deviatoric plane at low hydrostatic stress. Figure 2.12 suggests that this Mohr–Coulomb failure surface agrees reasonably well with experimental data along the compressive meridian but less well along the tensile meridian.

Post-yield stress–strain relationships

Various theories have been proposed for post-yield behaviour: perfect plasticity; isotropic work-hardening (or softening); kinematic work-hardening or a combination of isotropic and kinematic hardening. Kinematic hardening is only really needed for cyclic loading in concrete (Chen 1982). Most of the models (see Appendix C) assume isotropic hardening up to a peak stress and perfect plasticity thereafter. However, in experiments it has been observed that stress decreases with increasing strain after a peak value (see Figure 2.5). The shape of this descending branch of the stress–strain curve depends heavily on the confinement and the boundary conditions imposed by the experimental equipment (see Section 2.2.4). It is usually assumed that the hardening behaviour does not vary with age.

In classical plasticity models, the plastic strains vectors are obtained from the plastic potential and a flow rule, which can be either associated or non-associated. Associated flow rules assume that the plastic potential coincides with the yield function. Hence the vectors of plastic strain are normal to the yield surface. In the absence of experimental evidence to support a particular non-associated flow rule, the assumption of an associated flow rule is a common simplification (Chen 1982; Hellmich et al. 1999b).

Tension

In the plasticity models used in tunnel analyses, the Rankine criterion is generally used for yield in tension (see Appendix C). According to this brittle fracture occurs when the maximum principal stress reaches a value equal to the tensile strength (Chen 1982). The tensile strength is usually estimated from the compressive strength, using relationships for normal cast *in-situ* concrete. Post-failure behaviour in tension will be discussed later in Section 5.4.

Compression and tension

For states in which one of more of the principal stresses is compressive and the others are tensile, an assumption must be made about the nature of the yield surface. It is usually assumed that the presence of tension reduces the compressive strength linearly (see Figure 2.6, Chen 1982). However, there is some doubt about the exact effect (Feenstra & de Borst 1993).

In multi-surface plasticity models (e.g. Meschke 1996, Lackner 1995), a check is performed in each principal stress direction to see which of the yield surfaces is active, and the relevant yield criteria are then applied.

In a plasticity model, the material is assumed to behave in a linear elastic manner until the yield point is reached. Beyond that point, the stress increases (or decreases) in accordance with a hardening (or softening) rule relating the cohesion to the plastic strain up to the peak plastic strain. The plastic strains occur according to a flow rule in addition to the elastic strains. In the generalised case, the yield point becomes a surface in 3D stress space (e.g. see Figure 5.5) and is usually defined in terms of stress invariants (Chen 1982). The ratio of yield strength, f_{cy}, to ultimate strength, f_{cu}, (f_{cy}/f_{cu}) is about 0.40.

Figure 5.6 shows the variation of peak strains (i.e. strain at peak stress) with age, t. There is considerable scatter. However, a possible relationship between age in hours and peak strain in % is:

$$\varepsilon_{peak} = -0.4142 \cdot \ln(t) + 3.1213 \tag{5.7}$$

Figure 5.7 shows the ultimate peak strain plotted against age. As with Figure 5.6, the data has been extracted from published laboratory test results. All the values are considerably larger than the 0.35% limit stated in design codes (e.g. BS8110 1997, Eurocode 2 2004). This supports the view of early pioneers that sprayed concrete can withstand large strains at young ages. Meschke (1996) proposed a quadratic hardening law is used to calculate the change in cohesion (see Figure 5.8). This agrees well with the idealised stress–strain curve proposed in BS8110 Part 2 (1985).[8]

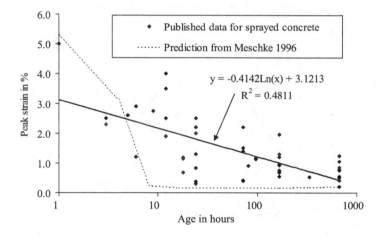

Figure 5.6 Peak compressive strain vs. age.

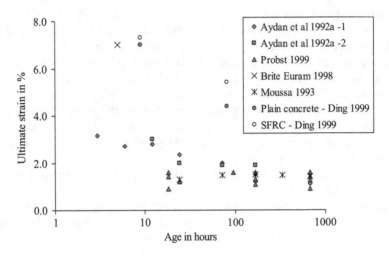

Figure 5.7 Ultimate compressive strain vs. age.

$$f = f_{cy} + 2\left(f_{cu} - f_{cy}\right) \cdot \left(\frac{\varepsilon_{pl}}{\varepsilon_{pl.\text{peak}}}\right) - \left(f_{cu} - f_{cy}\right) \cdot \left(\frac{\varepsilon_{pl}}{\varepsilon_{pl.\text{peak}}}\right)^2 \qquad (5.8)$$

where f is the cohesion (or strength), f_{cy} is the yield strength, f_{cu} is the ultimate strength, ε_{pl} is the current plastic strain and $\varepsilon_{pl\,\text{peak}}$ is the plastic strain at ultimate strength. The cohesion (on the compression meridian – Lode angle = 60°) can be related to the uniaxial strength, f_c, by equation 5.9 taking $\phi = 37.43°$.[9]

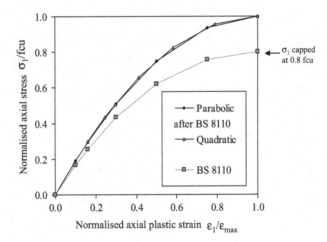

Figure 5.8 Theoretical strain hardening curves.

$$\text{cohesion} = f_c \cdot \frac{(1 - \sin\phi)}{2.\cos\phi} \tag{5.9}$$

Influence on the predictions of numerical models

Typically strain-hardening plasticity models predict an increase of 15–30% in the magnitude of deformations and a reduction of 10–25% in the magnitude of bending moments in the concrete shell, compared to an age-dependent elastic model (Moussa 1993, Hafez 1995, Hellmich et al. 1999c, Thomas 2003) – e.g. see Figure 5.1 and Figure 5.2. However, the influence of plasticity obviously depends on how heavily the lining is loaded and the ability of the ground to sustain the stress, which is redistributed back into it. If the ground is close to failure, the stress redistribution due to plastic deformation in the lining may exacerbate the situation (Hafez 1995).

5.4 TENSILE STRENGTH

Since the tensile strength of concrete is much lower than the compressive strength, in many normal load cases, failure in tension (i.e. cracking) may well occur while the compressive stresses are well below failure levels (Chen 1982). While some *in-situ* investigations have revealed tensile stresses in sprayed concrete tunnel linings (Hughes 1996, Negro et al. 1998) and cracking is of major concern, when considering permanent sprayed concrete linings, only the simplest tensile models have usually been used in design analyses. Namely, the concrete behaves in a linear elastic manner up to a tensile cut-off at the uniaxial tensile strength. This approach may simply have been adopted due to limits on computing power. The sections below include information on more sophisticated constitutive models even though they are not used in current design practice.

5.4.1 Unreinforced sprayed concrete

The tensile strength of unreinforced sprayed concrete, f_{tu}, is rarely tested. In the absence of other data, the tensile strength could be estimated from the compressive strength, using relationships for normal cast *in-situ* concrete, e.g. (Neville 1995):

$$f_{tu} = 0.30 . f_{cu}^{0.67} \tag{5.10}$$

The maximum principal stress failure criterion is the most commonly used. According to it, once the tensile stress acting on a plane exceeds the strength, a crack is formed and the stress carried across the crack falls to zero. In reality, if the width of the crack is not too great, 40 to 60% of the shear forces can still be carried due to aggregate interlocking (Chen 1982).

Therefore, the behaviour of concrete after cracking is highly nonlinear and orthotropic (Kullaa 1997). Alternatively, one can assume a maximum principal strain criterion, where cracks form once a limiting strain value is exceeded (Chen 1982).

A more sophisticated approach is to assume that the tensile stress of plain concrete decreases linearly, bilinearly or exponentially with increasing strain (Lackner 1995). The difficulty in determining the softening curve has led to the use of fracture energy to calculate the parameters for use in these models (Feenstra & de Borst 1993, Sjolander et al. 2018). The fracture energy is the area under a stress-deformation curve, and a correction is required to account for the size of the mesh elements. This approach can also be used for the post peak softening behaviour in compression too (Feenstra & de Borst 1993).

Meschke (1996) and Kropik (1994) used a (Rankine) maximum stress failure criterion followed by linear tension softening. The gradient of the descending stress–strain line is assumed to be E/100 – one-hundredth of the initial elastic modulus. While this would seem to model correctly the pre-crack and post-crack behaviour of plain concrete, the composite material of reinforced concrete actually exhibits tension hardening.

The cracks can be modelled either discretely or smeared over the elements in question. The concept of the smeared crack can be further subdivided into nonlinear elastic, plastic or damage theory models (Lackner 1995). Smeared cracks can be either fixed (once they have formed) or rotate their orientation as the direction of the tensile stresses changes. To avoid re-meshing, the smeared crack approach is usually adopted (Kullaa 1997).

5.4.2 Reinforced sprayed concrete

Like cracks, bar reinforcement can be modelled discretely (Kullaa 1997, Eierle & Schikora 1999). However, for both features, this can be a laborious process even in a simple 2D mesh and is too complex an approach at this time for the 3D analysis of tunnels. Reinforcement and the tensile behaviour of reinforced concrete are rarely simulated in numerical models, either explicitly (e.g. Haugeneder et al. 1990, Thomas 2003) or implicitly.

In the case of a smeared crack model, the reinforcing effects of steel and fibres may be incorporated by modifying the post-crack (tension softening) properties of the concrete elements. For example, it can be assumed that a fraction of the tensile stress across the crack can still be sustained. In the case of steel fibre reinforced concrete, various values have been proposed for this, ranging from 0.3 (Brite Euram C2 1997) to 0.35 to 0.5 (the minimum requirements for structural fibres according to fib 2010 – see Section 4.2.5). Obviously, a single fixed value does not take account of the variation in behaviour with crack width or the anisotropic distribution of the fibres. Moussa (1993) chose to multiply the value of the ultimate tensile strain by a factor of ten to account for the presence of reinforcement.

The fracture energy method described in Section 5.4.1 is often used when modelling fibre reinforced concrete in general – and sometimes in the design of fibre reinforced precast concrete tunnel segments to check crack widths and overall stability (ITAtech 2016). Although FRS is often used in SCL tunnels, the behaviour in tension is very rarely modelled.

Influence on the predictions of numerical models

Given the dearth of information on this subject, it is not possible to comment in detail on the influence of the model of tensile behaviour. In general, one may note that the assumption of an infinite tensile capacity (e.g. in an elastic model) obviously overestimates the capacity of the lining, while a brittle tensile cut-off would underestimate the capacity and result in an overestimate of stress redistribution. A pragmatic approach, which could be adopted, is to use a simple tension model and to compare the predictions of tensile stresses with the tensile capacity. More complex analyses can then be undertaken if required.

As part of a broader study Thomas (2003) modelled a lattice girder in a ring of sprayed concrete using cable elements for the three main steel bars. The presence of the lattice girder did not have any significant influence on the results of the numerical model due to its small area relative to a 1 m width of the concrete lining, which is the typical spacing of lattice girders.

5.5 SHRINKAGE

Figure 5.9 shows the variation of shrinkage strain with age. Given the scatter in experimental data, it appears that the simple ACI equation, with the constant B = 20 days and an ultimate shrinkage strain, ε_{shr00}, of 0.1%, may be used to predict the development of shrinkage with age (ACI 209R 1992) as a reasonable first approximation. Obviously, the value of B may vary depending on the characteristics of each mix (e.g. Jones (2007) quotes a value of B = 55).

$$\varepsilon_{shr} = \frac{\varepsilon_{shr\infty} \cdot t}{(B + t)} \quad \text{where } t \text{ is age in days} \tag{5.11}$$

Shrinkage is rarely modelled in design analyses on the assumption that the effects are much smaller than the ground loads. However, this may not be the case in lightly loaded linings (e.g. Jones 2007). Hellmich et al. (2000) found that in one set of 2D analyses bending moments were reduced by shrinkage but axial forces were relatively unaffected while in another example (back analysis of the Sieberg Tunnel) their model predicted a large reduction in hoop forces due to shrinkage.

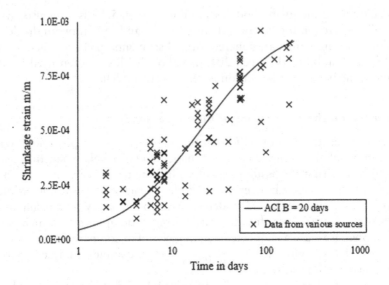

Figure 5.9 Shrinkage of sprayed concrete.

Data from Abler1992, Cornejo-Malm 1995, Ding 1998, Golser et al. 1989, Schmidt et al. 1987, Pichler 1994, Rathmair 1997, Zhgondi et al. 2018.

5.6 CREEP MODELS

Traditionally the high creep capacity of sprayed concrete has been hailed as a great benefit since this can dissipate stress concentrations and avoid overloading. While it may be very important in high stress environments, in shallow soft ground, it may be less important than the phenomenon of arching in the ground.

Rheology is the study of flow. Hence the term "rheological" is often used to cover empirically based creep models, such as those that have been idealised as an arrangement of simple units, each with certain defined behaviour (Neville et al. 1983). In the following sections other creep models – such as power law models – will also be discussed, with some comments on the effect of incorporating creep into the numerical model.

5.6.1 Rheological models

Typically, these models consist of Hookeian springs, Newtonian dashpots and St Venant plastic elements, arranged in series or parallel (Jaeger & Cook 1979, Neville et al. 1983), although more exotic units have been devised (e.g. springs in dashpots or Power's sorption elements). Figure 5.10 (a, b and c) shows the three most commonly used rheological models, in the analysis of sprayed concrete – namely, the generalised Kelvin (Voigt) model,

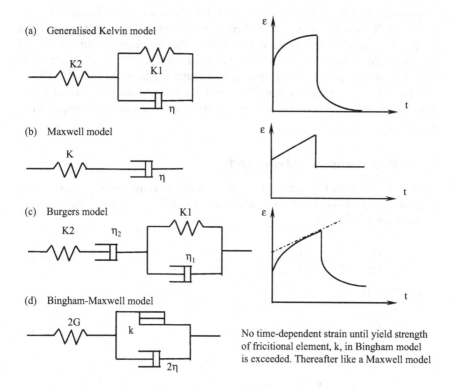

(a) Generalised Kelvin model

K2

K1

η

(b) Maxwell model

K

η

(c) Burgers model

K2 η_2 K1

η_1

(d) Bingham-Maxwell model

2G

k

2η

No time-dependent strain until yield strength of fricitional element, k, in Bingham model is exceeded. Thereafter like a Maxwell model

Figure 5.10 Rheological models.

the Maxwell model and the Burgers model, which consists of a Maxwell model in series with a Kelvin model.

The three models listed above are visco-elastic models, and so the principle of superposition can be applied. The spring stiffnesses and dashpot viscosities can be either linear or nonlinear. From Figure 5.10 (a), it can be seen that the Kelvin model produces a complete recovery on unloading, and so it is often used for fully recoverable transient creep. The Maxwell model produces no recovery and is used for steady-state creep, as well as stress relaxation, unlike the Kelvin model.[10] When combined in a Burgers model, one could say that they cover the whole of concrete's creep behaviour, with the Kelvin model replicating young concrete's behaviour and the Maxwell model the mature concrete behaviour. However, it would be unreasonable to expect that such a simple model could cover such complex behaviour, and more elaborate rheological models (e.g. Freudenthal–Roll model with one Maxwell and three Kelvin elements in series) have been proposed (see Neville et al. 1983, Chapter 14 for an overview).

Appendix D contains a list of rheological models used in analyses of sprayed concrete linings, together with their parameters. In addition to the generalised Kelvin model and Burgers model, a modified Burgers model (with two Kelvin elements) and a Bingham model have been proposed. All of the rheological models appear to have been formulated for deviatoric stresses only, although there is some experimental evidence to suggest that considerable creep may occur under hydrostatic loading too (Neville 1995). The models are almost exclusively based on the results of uniaxial creep tests. In addition to creep in the direction of loading, lateral creep occurs (see Section 2.2.7).

5.6.2 Generalised Kelvin model

In its mathematical formulation, the Kelvin model requires two parameters, G_k and η, in addition to the normal elastic moduli, as shown below.

For a uniaxial case:

$$\varepsilon_{xx} = \frac{\sigma_{xx}}{9.K} + \frac{\sigma_{xx}}{3.G} + \frac{\sigma_{xx}}{3.Gk}\left(1 - e^{-Gk.t/\eta}\right) \tag{5.12}$$

and in the 3D case:

$$\dot{e}_{ij} = \frac{\dot{S}_{ij}}{2.G} + \frac{S_{ij}}{2} \cdot \frac{1}{\eta_k}\left(1 - e^{-Gkt/\eta}\right) \tag{5.13}$$

$$\text{or} \quad \dot{e}_{ij} = \frac{\dot{S}_{ij}}{2.G} + \frac{S_{ij}}{2\eta_k}\left(1 - \frac{2Gk.e_{ij}}{S_{ij}}\right) \tag{5.14}$$

where ε_{xx} and σ_{xx} are the strain and stress in the x direction, K is the bulk elastic modulus, G is the shear elastic modulus, t is time, S_{ij} is the deviatoric stress, $\dot{s}_{ij}$ is the deviatoric stress rate and $\dot{e}_{ij}$ is the deviatoric strain rate. As noted above, creep is generally assumed to occur under deviatoric loading only (Jaeger & Cook 1979, Neville et al. 1983).

Research in the mid-1980s by Rokahr and Lux made a significant contribution to the understanding of creep effects in early age sprayed concrete. Based on the results of creep experiments on samples, they proposed a generalised Kelvin model, valid for ages from eight hours to ten days. They were able to model numerically what Rabcewicz had intuitively deduced – namely, that creep in the sprayed concrete reduces the stresses in the lining (Rokahr & Lux 1987) (see Figure 5.11). While the utilisation factor (stress–strength ratio) may be high at early ages, as the stress reduces due to creep and the strength increases, the utilisation factor falls and the factor of safety increases. Some subsequent fieldwork has supported this (e.g. Schubert 1988). Although a simple model, both the spring and dashpot parameters are stress dependent and the viscosity increases with age (see Appendix D).

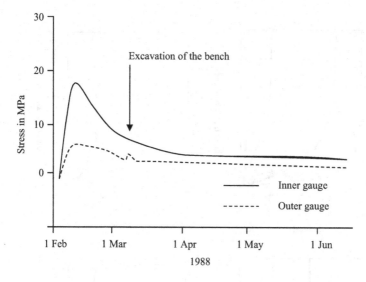

Figure 5.11 Stress reduction due to creep, computed from strain gauge data (Golser et al. 1989).

Others too (Swoboda & Wagner 1993, Kuwajima 1999, Sercombe et al. 2000) have adopted a generalised Kelvin model, most notably Kuwajima, who investigated how the parameters vary with age over the first 100 hours. However, for the purposes of back-analysing the creep tests, he chose average values, which leads to overestimates of strains. It was suggested that calculations could be simplified further, by assuming that the entire visco-elastic strain occurs instantaneously, without affecting the predictions adversely.

One of the weaknesses of research to date is its fragmented nature. Individual researchers have proposed theoretical models to suit the (limited) experimental data that they have been available to collect themselves. It would be preferable to calibrate the models against as much of the existing data as possible.

Figure 5.12 (a) to (f) shows over 200 estimates of specific creep strains versus age from seven data sets, presented in groups according to the age at loading (Thomas 2003). Interpretation of creep tests is complicated by the fact that often the loads were applied incrementally at different ages. The total creep strain due to one load increment may not have developed before the next was applied. Furthermore, the utilisation factor may vary considerably during the test due to the ageing of the material (Huber 1991). From Figure 5.12, age is clearly a very important influence on specific creep strain (which is the creep strain divided by the magnitude of the load increment).

The two creep parameters for the Kelvin model can also be described as the specific creep strain increment, $\Delta e_{ij\,\infty} = 1/(2G_k)$, and the relaxation

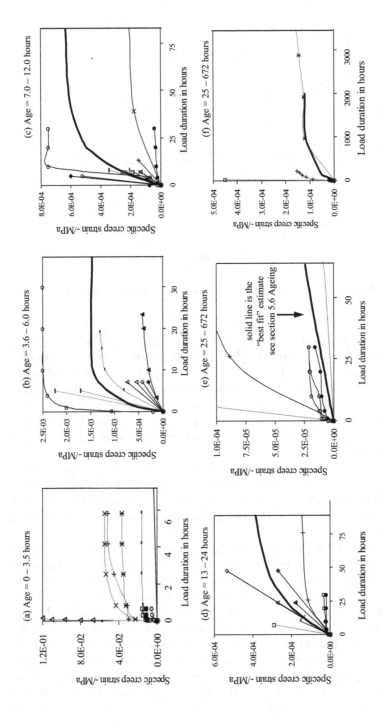

Figure 5.12 Specific creep strain of sprayed concrete, loaded at different ages.

Data from: Brite Euram B6 1997; Ding 1998; Huber 1991; Kuwajima 1999; Probst 1999; Rokahr & Lux 1987; Schmidt et al. 1987.

Table 5.3 Specific creep strain increment, $\Delta\varepsilon_{xx\,\infty}$, in -/MPa

Age at loading in hours	Lower bound	Average	Upper bound
0–3.5	1.5e-3	4.0e-2	8.0e-2
3.6–6.0	5.0e-4	1.5e-3	2.5e-3
7–12	2.0e-4	6.5e-4	7.5e-3
13–24	2.0e-5	4.0e-4	8.0e-4
25–672	5.0e-6	1.0e-4	2.0e-4

time, B, where $B = \eta_k/G_k$. In the formulation of Equation 5.13 the physical significance is clear. Namely, B is the time taken for 63.2% of the increment in creep strain to occur. When $t' = B$, $\Delta e_{ij\,\infty}.(2G_k) = 0.632 = 1 - e^{-t'/B}$ or when $t = 3B$, $\Delta e_{ij\,\infty}.(2G_k) = 0.95$. Table 5.3 and Table 5.4 summarise the data from Figure 5.12 (a) to (f) in terms of these parameters.[11]

Ageing

Yin (1996) proposed formulae of the form, $X = X_{28} \cdot a \cdot e^{c/t^{0.6}}/2(1+\upsilon)$, in line with established equations for predicting the development of stiffness and strength with age (e.g. Chang & Stille 1993) to account for the ageing of creep behaviour. The parameters for a Kelvin model could be assumed to vary with age in this fashion, according to the equations below.

$$\eta_k = \frac{1.5 \cdot e^{11} \cdot 1.0 \cdot e^{\left(-1.5/\left(\frac{T}{24}\right)^{0.6}\right)}}{2(1+\upsilon)} \quad \text{in kPa.s} \tag{5.15}$$

$$Gk = \frac{8.0 \cdot e^6 \cdot 1.0 \cdot e^{\left(-1.0/\left(\frac{T}{24}\right)^{0.4}\right)}}{2(1+\upsilon)} \quad \text{in kPa} \tag{5.16}$$

where T is the age of the sprayed concrete in hours, υ is Poisson's ratio and the other parameters have been chosen to obtain a reasonable fit to the data (see Figure 5.13 and Figure 5.14). The solid lines on Figure 5.12 (a) to (f) show

Table 5.4 Relaxation time, B, in hours

Age at loading in hours	Lower bound	Average	Upper bound
0–3.5	0.50	0.75	1.00
3.6–6.0	1.0	3.0	10
7–12	5.0	14.0	25
13–24	10	30	50
25–672	100	500	1,000

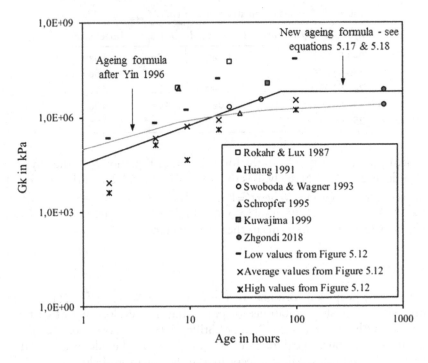

Figure 5.13 Shear stiffness (of spring in Kelvin rheological model), Gk, vs. age.

Predicted values from equations proposed by the authors listed in the key above as well as estimated values from the experimental data in Figure 5.12.

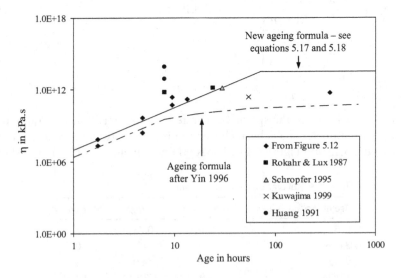

Figure 5.14 Viscosity of damper (in Kelvin rheological model), ηk, vs. age.

Predicted values from equations proposed by the authors listed in the key above as well as estimated values from the experimental data in Figure 5.12.

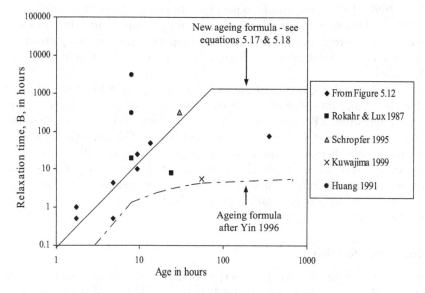

Figure 5.15 Relaxation time, B, vs. age (B = ηk/Gk).

the predicted specific creep strains from Equations 5.15 and 5.16. Figure 5.15 shows the relaxation time calculated from data in Figure 5.12 (a) to (f), along with the prediction from Equations 5.15, 5.16, 5.17 and 5.18.

Also plotted on Figure 5.13 and Figure 5.14 are approximate lines of "best fit" through the data:

$$\log_{10} Gk = a_g.\log_{10} T + b_g \tag{5.17}$$

$$\log_{10} \eta_k = a_\eta.\log_{10} T + b_\eta \tag{5.18}$$

where $a_g = 1.25$, $b_g = 4.50$, $a_\eta = 3.50$ and $b_\eta = 7.00$.

According to these equations, if T = 100 hours, $G_k = 1.0e^7$ kPa, which equates to a uniaxial specific creep strain, $\Delta\varepsilon_{xx\infty}$, of $3.33e^{-5}$ -/MPa (from Equation 5.12), and B = 2,780. The modified "Yin" formula predicts quite different values of $G_k = 1.89e^6$ kPa (equivalent to $1.76e^{-4}$ -/MPa) and B = 5. Experimental data suggests that the specific creep strain is about $1.0e^{-4}$ -/MPa and B = 500 (see Table 5.3 and Table 5.4). One should bear in mind that the majority of the data points for the loading age range of 25 to 672 hours refer to tests that were started at ages less than 80 hours. Hence the results may be biased towards a more pronounced creep behaviour than if more tests had started later.

Given the scant data for loading at ages greater than 100 hours, it would seem reasonable to assume that the sprayed concrete obeys the existing predictions for the creep of mature concrete. According to the ACI method

(ACI 209R (1992)), one would expect specific creep strains of about $1.08e^{-4}$ -/MPa and $0.68e^{-4}$ -/MPa after 700 hours for loading at ages 168 and 672 hours respectively. Eurocode 2 (2004) suggests a specific creep strain of about $1.11e^{-4}$ -/MPa after 700 hours, for normal C25 concrete loaded at 168 hours.

From a visual inspection of Figure 5.13 and Figure 5.14, it would seem that the ageing formulae after Yin overestimate how fast the creep occurs (i.e. underestimate B – see Figure 5.15) at all ages and the magnitude of the creep increment for ages greater than 100 hours. The logarithmic ageing formulae agree better, except in the age range greater than 100 hours where the creep increment is probably underestimated and the relaxation time overestimated. Therefore, these formulae are proposed as a good approximation, with the proviso that they are capped at 72 hours[12] (as shown on the figures).

Loading/unloading

In the case of varying loads, superposition is normally assumed (Neville et al. 1983). Unloading can be modelled as the addition of a negative load increment.

In terms of simulating creep in numerical models, the treatment of loading and unloading can be problematic. If the duration of each load increment is longer than 3B, each load increment could be treated separately, since the creep due to that increment would be essentially complete before the next one is added. If the load duration is shorter, one is faced with the question of whether to use the total stress in a zone/element or the stress increment during the creep calculation for each advance. Applying the normal principle of superposition would be very complicated. Even if one assumed that the creep strain increment during the advance of the tunnel face could be calculated individually for each load increment due to all the previous advance lengths, this would overlook the fact that, in this soil–structure interaction problem, the applied stress is not necessarily constant during the duration of an advance.

As a compromise one approach would be to assume that, when the time duration of each advance is greater than 1.5 B[13], only the increment in stress for that advance is used in the creep calculation (Thomas 2003). If the duration is less, it could be assumed that the time is insufficient for most of the creep strain increment to occur and so the increment in stress is allowed to accumulate.

Stress dependency

It has been widely reported that the creep strain rate is more than directly proportional to the applied stress for stresses greater than $0.5 f_{cu}$ (Rokahr & Lux 1987, Pöttler 1990, Aldrian 1991). A stress dependence is discernible in Figure 5.16 but, due to the large scatter, its exact nature is unclear. The

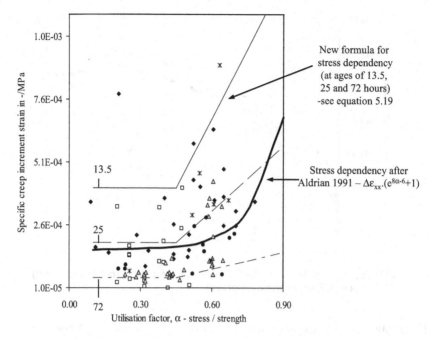

Figure 5.16 Specific creep strain increment vs. utilisation factor.
Data from Ding 1998, Huber 1991, Schmidt et al. 1987, Probst 1999.

exponential stress dependence proposed by Aldrian would appear to over-estimate the dependence for values of utilisation factors, α, of 0.45 to 0.75 and underestimate it for $\alpha > 0.75$. Based on Figure 5.16, as an initial estimate of the stress dependence, the following relationship for predicting the specific creep increment has been proposed for $\alpha > 0.45$ (Thomas 2003).

$$\Delta e_{ij \cdot \infty} = \frac{\left(1 + 2.5^{*}(\alpha - 0.45)/0.55\right)}{(2Gk)} \tag{5.19}$$

Validation

In the preceding sections, new relationships have been described for the parameters of a generalised Kelvin model. Before any new constitutive model is tried out in a full-scale model of a tunnel, it is important to validate it against the original experimental data it is based on. In this case, the creep model above was used to simulate the uniaxial creep test by Huber (1991) – see Thomas (2003) for full details. The model (with parameters chosen to match the test data – "VE matched") agrees to within –5% of both the analytical solution and the experimental data – see Figure 5.17.

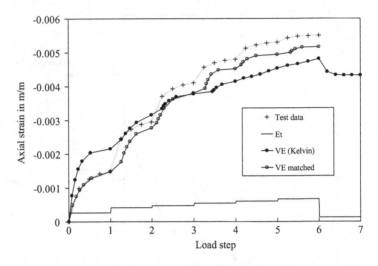

Figure 5.17 Comparison between FLAC models and test data from Huber (1991) (see Appendix E for explanation of key).

Encouragingly, Figure 5.17 also shows that the model (with the average parameters based on results of different researchers – "VE Kelvin") agrees reasonably well with the experimental data.

5.6.3 Burgers model

Several researchers have proposed using a Burgers model (see Figure 5.10 (c)). Based on a Burgers model but formulated as a time-hardening model, Petersen's model was valid for low levels of stress only (Yin 1996). Pöttler (1990) corrected this and expressed it in polynomial form. However, doubts remain over the validity of the parameters. The original data came from tests performed at ages greater than 30 hours. The creep rate calculated with Pöttler's formulae initially decreases with time but then rises after 1.5 days, in contrast to the observed behaviour. Yin attempted to correct this by re-formulating the model (as a power law creep model) and estimating the parameters, based on the assumption that they increase with age in the same manner as strength and elastic modulus, as proposed by Weber (see Appendix A and Figure 2.7 and Figure 2.14).

Zheng (referred to in Yin 1996) and Huang (1991) utilised an expanded Burgers model, though, in contrast to Neville et al. (1983), they viewed the Maxwell element as representative of the behaviour of young concrete and the Kelvin elements of mature concrete. While the irreversible viscous flow of a dashpot may make sense for very young sprayed concrete (or concrete in the case of Zheng, who was examining an extruded concrete lining), creep at mature ages is not fully reversible, as the Kelvin element implies.

5.6.4 Visco-plastic model

In the course of the Brite Euram project on sprayed concrete, a Bingham model for shear stresses was proposed for very young sprayed concrete – i.e. between two and seven hours old (see Figure 5.10 (d)). Four of the five parameters in this visco-plastic model are believed to vary with age, although this was not incorporated into the formulae. The fifth, Poisson's ratio, may also vary with age during the first 12 hours (see Figure 2.15). The resistance of the frictional element (St Venant element) increases with increasing hydrostatic stress. This model may have academic merit but for real tunnels it is not useful because it refers only to the earliest ages of sprayed concrete.

5.6.5 Rate of flow model

A model has been proposed for sprayed concrete, based on the rate of flow method (England & Illston 1965, Schubert 1988, Golser et al. 1989) with a view to back-calculating stresses from strain histories (Schropfer 1995). Strain histories can be broken down into four main components as below (see also Figure 5.18):

a) Instantaneous recoverable strain – elastic strain
b) Recoverable creep – delayed elastic strain
c) Irreversible ("yielding") strain – irreversible creep strain
d) Shrinkage (and thermal strain)

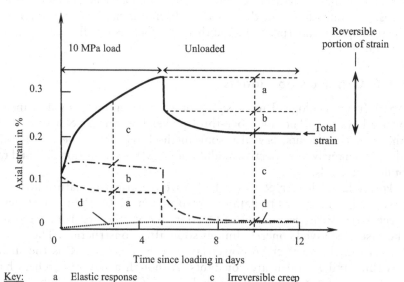

Key: a   Elastic response   c   Irreversible creep
        b   Delayed elastic response   d   Shrinkage

Figure 5.18 Decomposition of strains according to the Rate of Flow Method (after Golser et al. 1989).

Thermal strains were regarded as negligible (Golser et al. 1989). Based on extensive test results, equations have been written to describe each component (Golser et al. 1989) and subsequently refined to obtain better agreement at early ages and make the elastic modulus dependent on stress intensity (Aldrian 1991) – see Appendix D.

Aldrian (1991) proposed a relative deformation modulus, V^*, which is a reduction factor for the elastic modulus to account for the effect of the utilisation factor, α – the stress–strength ratio – as well as the age of the sprayed concrete (Appendix B). It is unclear what this actually means, since pre-loading, even to high utilisation factors, does not appear to reduce the initial slope of the stress–strain curve on reloading (Moussa 1993, Probst 1999). In contrast, the reloading modulus is usually higher. The factor, V^*, ranges from 1.0 at $\alpha = 0.0$ to 0.13 at $\alpha = 1.0$, at 28 days. It seems that this factor was intended for use in the incremental form of the equations to convert the initial elastic modulus into a tangent modulus and so account for the nonlinear nature of the stress–strain curve.

Many numerical analyses have been performed at the Montanuniversitat Leoben, in Austria, using the modified rate of flow method, most recently to investigate the transfer of loads between primary and secondary linings (and the ground) due to creep (Aldrian 1991, Rathmair 1997, Pichler 1994, Schiesser 1997). Golser and Kienberger (1997) contains an overview of this work. However, it has not been possible to implement this method in 3D finite element analyses (Rathmair 1997), and in 2D analyses it has been noted that agreement becomes poorer with increasing numbers of load steps. The rate of flow method has also been criticised because it relies on the principle of superposition, and therefore there is no allowance for plastic strains.

5.6.6 Other creep models

Apart from rheological models, the other existing creep models include power laws (e.g. Andrade's one-third power law (Jaeger & Cook 1979)) and creep coefficients, as well as some methods which have been mentioned already – namely, the effective modulus (or HME – see Section 5.2) and rate of flow methods.

Power laws (Jaeger & Cook 1979) are empirical in nature and were first used to fit curves to data on creep in metals. Of the three stages of creep – transient, secondary and tertiary – only the first is of interest in the case of sprayed concrete linings soon after construction. The general form for a transient creep power law is $\varepsilon = At^n$, where ε is the strain, t is time and A and n are constants. Although a few researchers have used forms of power laws in the analysis of SCL tunnels (Schubert 1988, Alkhiami 1995, Yin 1996, Rathmair 1997), they are not widely used because of their inferior ability to model the complex creep behaviour (e.g. the existence of recoverable and irrecoverable portions of creep).

Because of their widespread use in general engineering, power law creep models often come as standard in numerical analysis programs (e.g. ABAQUS, FLAC), but it is not always possible to combine the creep model with more sophisticated elastic or elastoplastic models in those programs (Rathmair 1997).

Standard methods for calculating the effects of creep in concrete have been published by various bodies, such as CEB-FIP and the American Concrete Institute (Neville et al. 1983, Chapters 12 and 13). Both use a creep coefficient, which is a combination of parameters that account for factors such as water–cement ratio, cement content and size of the structural member. While the strength of these methods lies in their ability to obtain a coefficient for a wide range of concretes and situations, their weakness lies in the fact that the parameters are valid for hardened (normal) concrete, aged seven days old or more. They cannot replicate the early age behaviour of sprayed concrete (Han 1995, Kuwajima 1999), and so they are generally not suitable for sprayed concrete. Furthermore, they may not be suitable for cases when variable loads are experienced by a tunnel lining. However, a creep coefficient can be incorporated in an ageing elastic model (see the *refined* HME model described in Section 5.1).

Influence on the predictions of numerical models

Creep of sprayed concrete has long been postulated as being responsible for easing stress concentrations in linings (Rabcewicz 1969, Rokahr & Lux 1987, Soliman et al. 1994). Many numerical studies (albeit in 2D) have been performed which have demonstrated significant reductions in axial forces and bending moments when a creep model is used for the lining (e.g. Huang 1991, Schropfer 1995, Yin 1996), but hard evidence from the field in support of this is scarce.

Since the load-bearing system is a composite consisting of the ground and the lining, movement of the lining should be expected to cause movement of the ground and a change in the load on the lining. Therefore, this case is not as simple as the standard laboratory creep test in which a constant load is applied to a sample and the increasing strain over time is recorded. Whether the creep of the lining results in a reduction in the lining stresses depends on the strength and material behaviour of the surrounding ground. If the ground is elastoplastic or susceptible to creep itself, the load in the lining might be expected to actually increase following creep (Pöttler 1990, Schiesser 1997, Yin 1996, Hellmich et al. 2000). The effects also depend on the rate of creep in the ground (Hellmich et al. 1999c). It is known that the creep capacity of concrete decreases rapidly with age. Several researchers have proposed relationships to predict this (e.g. Golser et al. 1989, Rokahr & Lux 1987, Pöttler 1990 Aldrian 1991), but there is little agreement between them on the rate of change of this property (see Figure 5.19 – normalised age-dependency of creep rate or specific creep). Similarly, some

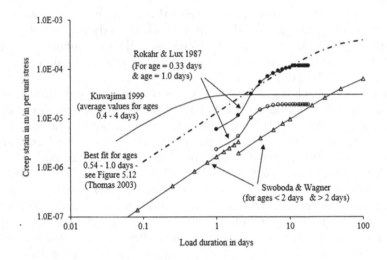

Figure 5.19 Predicted specific creep values.

have included stress dependency in their creep models (e.g. Golser et al. 1989, Aldrian 1991, Probst 1999).

A few numerical models, that have included creep, have shown unrealistically large reductions in stresses. For example, in analyses performed by Rathmair (1997) the axial force was reduced to 5% of its initial value, although the same author reports a reduction of stress of only 50% in laboratory relaxation tests on sprayed concrete. However, Thomas (2003) did find that, even in lightly loaded shallow tunnels, creep may lead to reductions in axial lining loads and bending moments – e.g. see Figure 5.1 and Figure 5.2. The extent of the reduction depends heavily on the parameters chosen for the creep model, ranging from 10 to 50%. Of the creep models investigated, the model described in Equations 5.17 to 5.19 was found to give the most realistic predictions. Therefore, it is recommended that the creep model is calibrated against laboratory data for the exact sprayed concrete mix planned for each individual project. This particular study also showed how creep can "smooth" out peaks in the stresses in linings – see Figure 5.20. As one would expect, creep also leads to higher predictions of lining deformations – see Figure 5.21.

5.7 AGEING

Ageing makes the task of modelling sprayed concrete considerably more complicated than is the case for other lining materials. While in a design calculation, one may assume that in a tunnel the load increment due to an advance is applied as soon as the ground is excavated and choose values of parameters (e.g. stiffness) that are consistent with the ages of the different

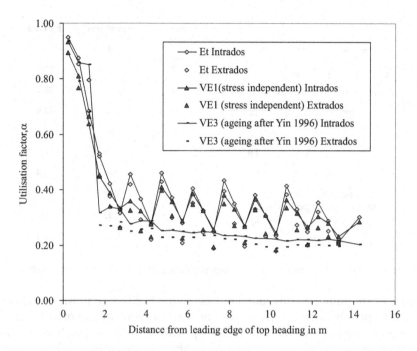

Figure 5.20 The effect of creep on utilisation factors.

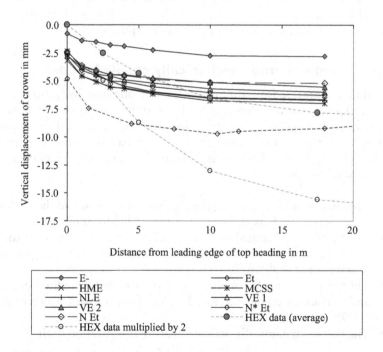

Figure 5.21 Crown displacement vs. distance from leading edge of top heading.

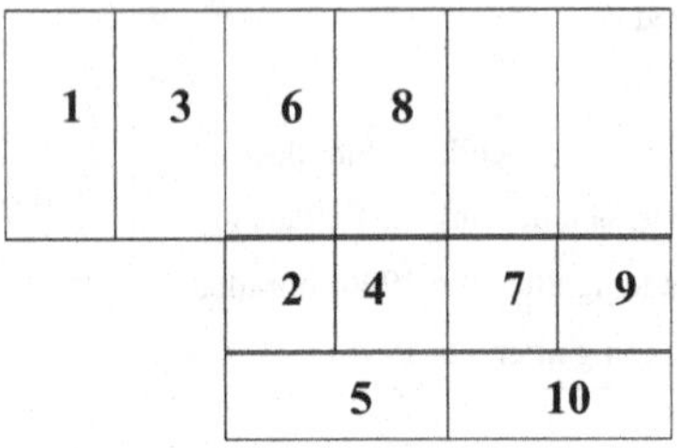

1	3	6	8		
		2	4	7	9
		5		10	

Schematic of the tunnel, showing
the stages of the top heading,
bench and invert excavation sequence

Stage	Age at start	E / GPa
1	0	0
2	24	17.0
3	48	20.2
4	72	21.6
5	96	22.4
6	120	23.0
7	144	23.4
8	168	23.7
9	192	23.9
10	216	24.1

Figure 5.22 Typical approximation of age-dependent stiffness in a numerical model.

parts of the lining at this moment (see Figure 5.22), one must also check that the new parameters are consistent with the constitutive model. All of the properties of sprayed concrete vary with age and Appendices A and B contains numerous empirical relationships for predicting the most commonly used properties (e.g. strength) at all ages. The age of the individual parts of the lining can be estimated, based on their distance from the tunnel face for a given advance rate and excavation sequence.

5.7.1 Thermo-chemo-mechanically coupled model

The fundamental reason for ageing is the ongoing hydration of the cement. Hence various researchers have sought to quantify how the degree of hydration (ξ) varies with time and then relate all the material properties to this. An overview of the theory can be found in Ulm and Coussy (1995, 1996). The thermo-chemo-mechanically coupled model aims to account for:

- The chemo-mechanical coupling between hydration and the evolution of properties such as strength, stiffness and autogenous shrinkage
- The thermo-mechanical couplings such as dilation due to the exothermic hydration reaction (Hellmich & Mang 1999) or damage criteria (Cervera et al. 1999b)
- The thermo-chemical couplings such as the reduction in final strengths and stiffnesses due to increased curing temperatures (Cervera et al. 1999b)
- The thermodynamically activated nature of hydration itself

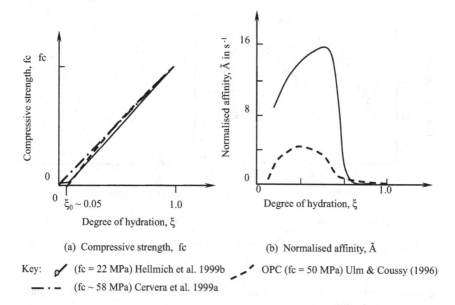

(a) Compressive strength, fc

(b) Normalised affinity, Ã

Key: ⟋ (fc = 22 MPa) Hellmich et al. 1999b   - - -' OPC (fc = 50 MPa) Ulm & Coussy (1996)

— · — (fc ~ 58 MPa) Cervera et al. 1999a

Figure 5.23 Hydration kinetics for sprayed concrete (Hellmich et al. 1999b).

An underlying intrinsic material function – the chemical affinity or driving force of hydration ($A_{(t)}$ or $A_{(\xi)}$) – can be determined experimentally (see Figure 5.23). Being intrinsic, it is meant to be independent of field variables and boundary effects (Hellmich 1999a, Hellmich et al. 2000). However, Figure 5.23 shows that there can be a large variation in the reported profiles of the normalised chemical affinity with most values closer to the line from Ulm and Coussy (1996) – see also Hellmich et al. (1999a) and Hellmich and Mang (1999). This is probably due to the fact that E_A/R also varies, depending on the mix and temperature. Reported values range from 4,000 to 5,000 K. Both parameters are needed to fully characterise the mix. Strength can then be related to the progress of the normalised chemical affinity function with time, t, as below.

$$\frac{d\xi}{dt} = \left(1 - \xi_0\right)\left[\frac{df_c/dt}{f_{c,\infty}}\right] = A_{(\xi)} \cdot e^{-E_A/RT_t} \tag{5.20}$$

$$A_{(\xi)} = a_A \cdot \left[\frac{1 - e^{-b_A \cdot \xi}}{1 + c_A \cdot \xi^{d_A}}\right] \tag{5.21}$$

The exponential term in Equation 5.20 accounts for the thermodynamically activated nature of hydration. T is temperature in K. ξ_0 is the "percolation

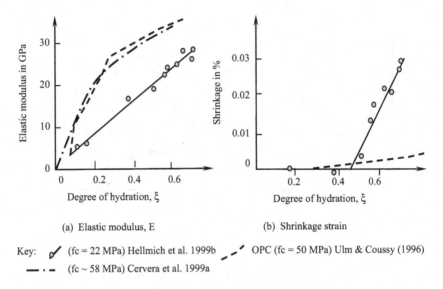

(a) Elastic modulus, E   (b) Shrinkage strain

Key: ⟋ (fc = 22 MPa) Hellmich et al. 1999b _ _ ⟋ OPC (fc = 50 MPa) Ulm & Coussy (1996)
      — · — (fc ~ 58 MPa) Cervera et al. 1999a

Figure 5.24 Variation of stiffness and shrinkage with the degree of hydration (Hellmich et al. 1999b).

threshold", below which the concrete cannot sustain any deviatoric stress and f_c is the compressive strength. a_A, b_A, c_A and d_A are all constants (see Appendix F). Similarly, other relationships have been proposed between properties such as tensile strength, autogenous shrinkage and stiffness and the degree of hydration (ξ) (see Figure 5.24 a and b) – see also Appendix A and Eierle & Schikora (1999). As noted earlier with the chemical affinity, there is a wide variation between reported results. Sercombe et al. (2000) contains tentative relationships[14] for the development of short- and long-term creep with the degree of hydration. In common with the empirical formulae, it is assumed that the hydration kinetics are independent of the loading history.

Alternatively, the effect of temperature on hydration can be accounted for by the simpler method of equivalent age (Cervera et al. 1999a, D'Aloia & Clement 1999). An elevated temperature speeds up hydration. The value at a given age can be read from a graph of the parameter's growth at a reference temperature, using an age, which has been corrected for the more advanced degree of hydration. Cervera et al. (1999a) note that normally all of the cement does not hydrate and the final degree of hydration can be found from:

$$\xi_{\infty} = 1.031 \, w/c / \left[0.194 + w/c \right] \sim 0.71 \text{ for } w/c = 0.43 \tag{5.22}$$

The initial "percolation threshold", ξ_0, (sometimes known as the critical degree of hydration) has also been found to vary with water–cement ratio (Byfors 1980):

$$\xi_0 = k.w/c \qquad\qquad (5.23)$$

where k varies between 0.40 and 0.46 and w/c is the initial water–cement ratio.

$A_{(\xi)}$ profiles have only been developed for a few concretes so there is limited data for some of the key parameters, e.g. the profile itself or the percolation threshold. Hence the parameters should be determined from laboratory tests for each specific mix for a project. Another means of determining $A_{(\xi)}$ is to work back from data on the development of strength with time. However, if one already knows this, other parameters (e.g. stiffness and tensile strength) can be estimated directly. The benefit of the thermo-chemo-mechanical approach is that it links the properties to the fundamental process behind them – namely, hydration. This provides some further insight into the behaviour of concrete at early ages and makes the inclusion of temperature effects straightforward where this is relevant. While temperature effects are important for massive structures (e.g. Aggoun & Torrenti 1996, Hrstka et al. 1999), most tunnel linings are quite slim (i.e. less than 400 mm thick) and the elevated temperatures due to the heat of hydration are short-lived. Figure 2.21 shows that the effect of elevated temperature on hydration is limited for the temperatures experienced in most SCL tunnels.

However, temperature effects may be discernible in lightly loaded tunnels. Jones (2007) coined the term "ground reaction temperature sensitivity" to describe the stresses induced in radial pressure cells in one shallow tunnel due to the expansion of the ring during hydration of the concrete. Even where temperature does not induce significant stresses compared to the ground load, it should be noted that temperature changes can induce fluctuations in measured stresses in SCL tunnels (see Section 7.3.1).

5.8 CONSTRUCTION SEQUENCE

Subdivision of the heading

For segmentally lined tunnels, it can be reasonably assumed that the lining is constructed in one action, albeit at a certain distance from the actual face. In SCL tunnels the excavation and construction sequence is much more complex yet this is not always replicated fully in numerical modelling. The way in which the sequence is simplified can have a great impact on the results. For example, modelling the excavation as full face rather than according to the exact sequence can lead to a more even pattern of lining loads (Guilloux et al. 1998) and less unloading in the ground (Minh 1999).

Advance length and rate

Experience and common sense suggest that increasing the advance length or rate may result in higher loading of the lining (Pöttler 1990). Convergence can increase by as much as 50% if the advance rate is increased from 2 m to 8 m per day, while the hoop load in the lining could fall by 15% for a tunnel in soft ground (Cosciotti et al. 2001). The difference reduces as the stiffness of the ground increases. The load decreases because the sprayed concrete is more heavily loaded at a young age and deforms more. This permits more stress redistribution in the ground and therefore a lower load in the final case (provided that the ground can sustain the load that has been redistributed back into it). However, there is the risk that the lining will be damaged through overstressing. Kropik (1994) reports a 20% increase in crown deformations when the advance length is increased from 1 m to 2 m in a soft ground tunnel. Hellmich et al. (1999c) noted the importance of when the lining acts as effective support in determining the final loads, despite the simplifications of the construction sequence in their numerical model. It was also noted that the sooner the lining can carry load (i.e. the sooner the percolation threshold is passed in the thermo-chemo-mechanical constitutive model – see Section 5.7.1), the higher the loads in the lining. This applies for both axial forces and bending moments.[15] The concept of delaying the installation of the completed lining in highly stressed rock tunnels, in order to reduce the load in the final lining, is a basic tenet of the NATM.

The case of soft ground is somewhat different. There is some evidence to suggest that in soft ground the load in the lining is actually lower if the lining is installed sooner rather than later (Jones 2005, Thomas 2003). This is not necessarily at odds with the NATM philosophy since it recognised that, beyond a certain point, delaying the installation of the lining results in higher loads (due to the "loosening" of the ground around the tunnel"). In soft ground, plastic deformation or strain softening could cause higher loads. For tunnels in stiff over-consolidated clays, Jones (2005) ascribed the increase in long-term load to the equalisation of negative pore pressures that had been generated by the unloading of the ground during construction of the tunnel. This occurred mainly in the invert. The greater the unloading (e.g. due to a delay in closing the invert), the larger the negative pore pressures and the higher the subsequent increase in load.

Similarly, in line with experience on site (e.g. Thomas et al. 1998), Kropik (1994) noted that closing the invert early (i.e. close to the face) led to a reduction in deformation of the lining.

In the past "unlined" analyses have often been used in 2D and 3D simulations (e.g. Gunn 1993, Krenn 1999, Minh 1999, Burd et al. 2000). Researchers have often reported that they obtained a good correlation between the results of such analyses and field data of settlement. Obviously, such simulations are completely unrealistic, and any apparent good match probably stems from peculiarities of the analyses (e.g. the use of a prescribed volume loss or a simple ground model) which prevent the failure that would occur in the real case. Dasari (1996) notes that in his work there was little difference between 2D analyses of lined and unlined tunnels but that the introduction of the lining greatly reduced the settlements in the 3D analyses of the same tunnel.

Load development

The load from the soft ground is generally assumed to increase with time monotonically (e.g. Grose & Eddie 1996). While this may be true for segmentally lined tunnels (e.g. Barratt et al. 1994), given the complex excavation sequences and geometries in SCL tunnels, the mode of action of the lining may well change and the loading may vary. For example, one could consider the lining in the top heading as cantilevering off the completed rings behind it initially, resulting in bending in the longitudinal direction (see Figure 1.1). When the ring is completed, the lining acts mainly in compression and the main bending moments act in the hoop direction.

It is worth noting in passing that the longitudinal stresses in tunnel linings have rarely been examined in detail. In a detailed numerical modelling study, Thomas (2003) found that in the top part of the lining – i.e. above axis – the extrados of the lining was in tension due the relative movement of the ground towards the face. The intrados was in compression. This process continued far back from the face, and therefore it was the stiffness here that helped to determine the longitudinal forces and bending moments in the final situation – see Figure 5.25 and Figure 5.26. Hence, the

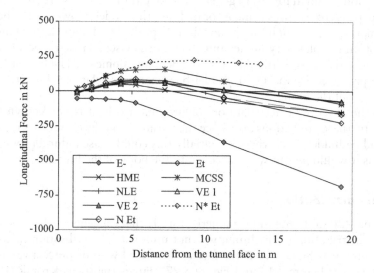

Figure 5.25 Longitudinal axial forces in the crown vs. distance to face.

NB: Compressive forces are positive / See Appendix F for explanation of key.

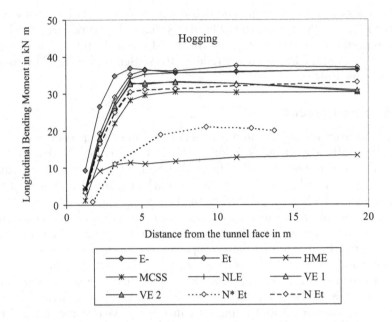

Figure 5.26 Longitudinal bending moments in the crown vs. distance to face.

See Appendix F for explanation of key.

constitutive model of the sprayed concrete has a large influence on the predicted longitudinal forces. In general, the longitudinal forces are small if the
age-dependent stiffness is incorporated into the model. However, the bending moments were high for all models, except the HME model. The tension
cut-off in the plasticity model and creep in the visco-elastic models helped
to reduce the tension and therefore the bending moments. The lowest loads
were predicted by the HME model because of its relatively low stiffness
throughout the lining.

In the absence of field data on longitudinal loads, it is not known how
realistic these predictions are. Since the tension is on the extrados, cracking
would be hidden from view. Potentially this could cause a durability problem as it would permit water access into the body of the lining.[16]

Stress distribution

Thomas (2003) found in his numerical study that, in general, the predicted
stress distribution in the lining was not uniform. The utilisation factor in
the lining was highest near the face and reduced with distance away from
the face (see Figure 4.13 and Figure 5.27). Furthermore, as Kropik (1994)
found, the stress at the leading edge of each advance length was found to be
greater than at the trailing edge. This may have been due to the difference

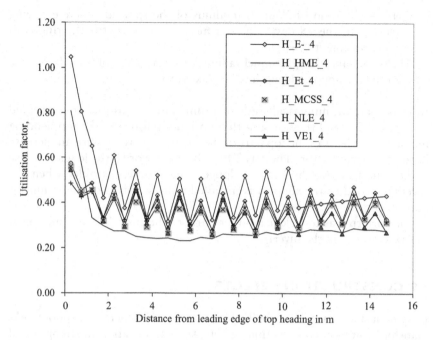

Figure 5.27 Utilisation factors in the crown at extrados vs. distance from leading edge.

in the concrete stiffness in adjacent parts of the lining and the higher radial loads at the leading edge. Stresses were higher in the first part of the lining that was constructed – i.e. the stresses were higher in the top heading than the Invert.

The Sequence Factor

Given the influence of the construction sequence in SCL tunnels, it would be useful to have a means of estimating the impact of the key parameters. Three of the main parameters – advance length (AL), advance rate (AR) and distance to ring closure (RCD) – are interrelated since altering one will affect the others. One way to assess the impact of changes in construction sequence may lie in considering the combination of those key parameters into a new factor – the "Sequence Factor":

$$\left(\frac{RCD}{AR} \cdot \frac{Ex}{E_{28}} \cdot \frac{AL}{R} \right) \tag{5.24}$$

which consists of:

RCD/AR – ring closure distance/advance rate ~ the time taken to close the ring

E_x/E_{28} – the ratio of Young's modulus of the sprayed concrete at ring closure to the 28-day value ~ a measure of how stiff the ring is in compression at closure

AL/R – advance length/tunnel radius (R) ~ a measure of the relative size of the unsupported length during excavation

Results from one numerical study of a tunnel in soft ground are presented in Figure 5.28, and they suggest that both hoop forces and hoop bending moments – at least in the top part of the tunnel – increase with an increase in the Sequence Factor (Thomas 2003). In other words, the loads increase if the time to close the invert is longer; the concrete is older when the invert is closed or the support is installed more slowly (i.e. longer unsupported length).

The pattern was less clear in other parts of the tunnel lining (such as at the axis level or in the invert).

5.9 CONSTRUCTION DEFECTS

Despite the fact that SCL tunnelling is known to be vulnerable to poor workmanship, common construction defects, such as variation in strength and quality, poorly constructed joints and variations in shape and thickness, are

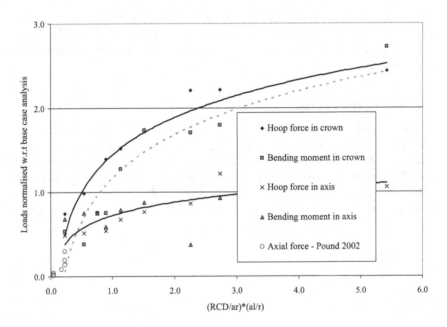

Figure 5.28 Normalised hoop loads vs. (RCD/ar) × (al/r) corrected for tunnel radius and stiffness at ring closure at 9 m from the face.

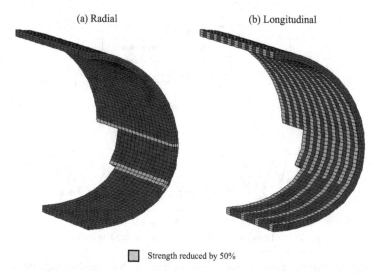

Figure 5.29 Locations of joints in mesh of the SCL tunnel modelled by Thomas (2003).

not normally considered in design calculations. In contrast, in the design of segmental linings, it is routine to consider the effects of ovalisation (due to ground deformation or poor build quality) and the misalignment of segments (so-called "stepping").

Stelzer and Golser (2002) examined the effects of the sort of variations in profile of a sprayed concrete lining that are found in drill and blast tunnels. In their detailed study of both small scale models and the back analysis of them with a numerical model, they found that imperfections could reduce the structural capacity of a lining by more than 50%. The imperfect linings tended to deform more too.

SCL tunnels contain many joints, and these can be areas of weakness (e.g. HSE 2000, Figure 3.16). In one study of an SCL tunnel in soft ground, the strength of the lining was reduced by 50% at the joints in the numerical model (see Figure 5.29). Figure 5.30 shows that the presence of weak zones at circumferential and radial joints can alter the stress distribution within the tunnel lining (Thomas 2003). Weak radial joints tended to increase the loads near the face while weak circumferential joints reduced the loads and the lining functioned more like a tube consisting of discrete rings. Weakening both radial and circumferential joints worsened the situation in the critical area near the face.

5.10 SUMMARY

Numerical modelling plays an increasingly important role in the design of SCL tunnels. The choice of constitutive model for the sprayed concrete can

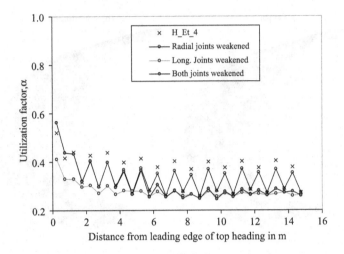

Figure 5.30 Utilisation factors in the crown vs. distance from leading edge (for models with weak joints).

have a significant influence on the results of this numerical modelling. This applies both to the tunnel lining itself and the ground around it. Figure 5.21 illustrates how predictions of lining deformations can vary depending on the model, while Figure 5.31 shows that even the surface settlement can be influenced. Despite the complex behaviour of sprayed concrete, simple models tend to be used.

Figure 5.32 demonstrates the potential benefit of using more sophisticated numerical models (Thomas 2003). These models predicted bending

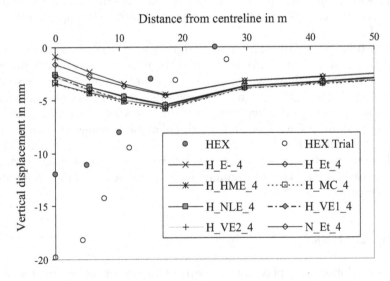

Figure 5.31 Transverse surface settlement profile at 18 m from the face (Thomas 2003).

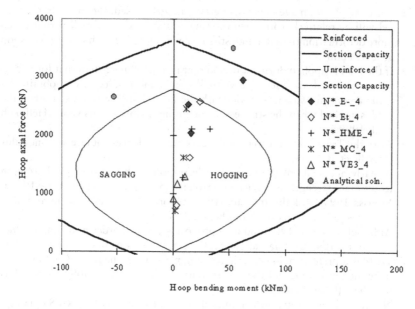

Figure 5.32 Results from numerical model of a shallow SCL tunnel in soft ground (Thomas 2003).

moments that would have been small enough to permit the use of steel fibres instead of wire mesh reinforcement. Having removed this major durability concern related to bar reinforcement, a "one pass" permanent sprayed concrete lining becomes a viable option. Taking one example, this could result in an estimated cost saving of 30%.[17] However, one should note that the lining loads are only one of many considerations in lining design.

There is no single right answer to the question of which constitutive model should be used for sprayed concrete. Above all else, the chosen model must be appropriate for the design calculation. At this point, it is worth reiterating that the constitutive model for the lining is only one part of the numerical model. Different aspects of the model for sprayed concrete may assume more or less importance depending on what the main focus of interest is. For example, the model of the tunnel lining may have little influence on the prediction of surface settlements. One should also bear in that, in principle at least, the same outcome could be obtained from a complex model by using (widely) different parameters and/or "sub-models" for the various elements.

NOTES

1. This section draws heavily on that thesis. Appendix E contains the key for the numerical models from that thesis which are referred to in figures in this chapter.

2. For example, if an age-dependent elastic model is used, the invert of the tunnel will be relatively soft when the ring is first closed and will permit much more deformation than if a high stiffness were assigned to that section as soon as it was built.

3. If the load is applied during a period that is much longer than that for hydration, numerical analyses give virtually the same results as one obtains by using a constant stiffness model with the 28-day stiffness, since most of the load is applied when the stiffness of the lining is close to this value (Hellmich et al. 1999c).

4. Also, the Trost–Bazant creep model uses an age-dependent effective modulus (Neville et al. 1983).

5. At early ages, due to the more ductile response, the ratio of yield strength to ultimate strength is higher – between 0.5 and 1.0 (Aydan et al. 1992a, Moussa 1993). Rokahr and Lux (1987) report a linear response up to 0.8 at 24 hours (see also Section 0 and Figure 2.5).

6. Although Moussa (1993) did also propose a seventh order polynomial function to describe the uniaxial compressive stress–strain curve more precisely.

7. There is a preference in geotechnical engineering for the name Coulomb and in applied mechanics for the name Mohr, hence the name Mohr–Coulomb is used here (Chen 1982).

8. NB: the maximum strength permitted by BS8110 Part 2 (1985) is 0.8 f_{cu} (= f_{cyl}), and this is only for the analysis of non-critical sections.

9. From Chen (1982); Yin (1996) proposed 40°.

10. Strictly speaking creep refers only to increasing strain with time, under a constant load. Relaxation refers to the reduction in stress over time observed in samples held under a constant strain.

11. The data come from uniaxial creep tests, and therefore the parameters have been determined using $\Delta\varepsilon_{xx\infty}$ = 1/3Gk.

12. Obviously, creep remains age-dependent beyond the age of 72 hours and the logarithmic ageing formula could be amended to reflect ageing in line with published formulae or data (e.g.: Eurocode 2 or ACI 209R).

13. 1.5 B = the time for 0.78 of the creep strain increment to occur.

14. The relationships are based on one creep test on a sprayed concrete sample at an age of 28 days.

15. Hellmich et al. (1992c) found that varying the stiffness of the lining, which was modelled with a linear elastic model of constant stiffness, made little difference to the hoop forces but influenced the hoop bending moments greatly. From this they concluded that the hoop force does not depend on the material properties of the lining. However, this is only true for the simplest cases in their study. When there is true interaction between the lining and the ground, the early age stiffness, and particularly the point at which it becomes effective, had a great influence on the stresses in the lining.

16. There is some anecdotal evidence from segmentally lined pilot tunnels that, during enlargement to the full cross-section, longitudinal movements in the ground can drag pilot tunnel linings into tension, causing the joints between rings to open.

17. Based on a rough estimate of the saving in time and materials cost for the HEX Platform tunnel, assuming that no secondary lining would have been required for the steel fibre reinforced sprayed concrete option (Thomas 2003).

Chapter 6

Detailed design

Notwithstanding the veracity of the old adage that "if it can't be designed on the back of a cigarette packet, it can't be built", detailed calculations are essential as justification and evidence of robust engineering in modern design practice. Sadly, they are rarely pocket-sized. This chapter provides guidance on detailed design of the sprayed concrete lining only, in a range of ground conditions and special cases. Topics such as face stability or rock bolt design are not covered. More details on tunnel design in general can be found in the following texts: Szechy (1973), Hoek & Brown (1980), BTS Lining Design Guide 2004) and Chapman et al. (2017).

When using analytical or numerical design tools, the procedure for detailed design calculations is generally the same. The loads in the lining are estimated. These loads are compared with the capacity of the lining, with normal partial safety factors applied, normally using a moment–force interaction diagram (e.g. see Eurocode 2 2004). The deformations (of the lining and ground) are also estimated, and these are used to determine trigger values, if required (see Section 7.3).

At the risk of appearing tedious, once again it should be stressed that the sprayed concrete is only one part of the support system for a tunnel. The soil–structure interaction must be modelled realistically. For a given tunnel, the modelling of the ground or the construction sequence may prove more critical to the design than the subtleties of the behaviour of the sprayed concrete.

6.1 DESIGN FOR TUNNELS IN SOFT GROUND

This section covers the design of sprayed concrete linings in soft ground (i.e. where the soil or weak rock behaves as a continuum). The key mechanisms of behaviour are plastic yielding or failure in the ground around the tunnel (see Table 3.2), and the sprayed concrete must provide immediate support. Face stability, along with the available equipment, drives the choice of excavation sequence.

6.1.1 Key behaviour of sprayed concrete

Because of the role of immediate support, the age- and time-dependent behaviour of the sprayed concrete may well be relevant. Excavation sequences tend to have multiple stages, and the intermediate load cases when the lining is incomplete should also be checked as they may be more critical than the long-term loading. Nonlinear behaviour may occur during these stages.

6.1.2 Determining the loading on the sprayed concrete

The behaviour of soft ground itself may be quite complex: *in-situ* horizontal stresses may exceed vertical stresses; K_0 may vary with depth; nonlinearity elastic/plastic behaviour and anisotropy may influence strongly the loads on the tunnel. Depending on the permeability of the ground, undrained behaviour may also feature. As an example, Addenbrooke (1996) and van der Berg (1999) describe the features relevant to one type of soft ground, London Clay.

Consequently, simple analytical methods alone may not be adequate, although they are often used in early stages of design – see Table 4.2 for examples. Numerical modelling may provide more realistic estimates of the ground loads. Due to the weakness of the ground, support is installed quickly, and hence the relaxation factor is low (see Table 4.1), often around 50%. Alternatively, a target volume loss can be used in the numerical model (i.e. the ground is permitted to relax until a certain amount of deformation has occurred). Good quality construction results in volume losses around 1.0%.

6.1.3 Lining design

Despite the comments above, a single SCL tunnel in soft ground should not present too many difficulties to a designer.

Some excavation sequences have sharp angles (see Figure 3.3), and design calculations tend to predict high bending moments there. In practice, often these do not appear to occur, most probably due to arching in the ground or creep within the sprayed concrete. One way to check how important these stress concentrations may be is to insert a "pin joint" (i.e. to release the fixity on rotation) and see how this changes the outcome of the analysis. Overall the shape of the tunnel should be fairly circular to minimise stress concentrations in the ground or the lining (see Figure 6.1). Sharp corners may increase the dead loads on the lining as the arching in the ground follows more gentle curves or induce high bending moments in the lining as noted above. Horse-shoe shaped cross-sections tend to experience instability under the footings. Increasing the bearing area of the footing with "elephant's feet" can combat this but in the end a more rounded cross-section and early ring closure may be the only way to prevent instability in the invert.

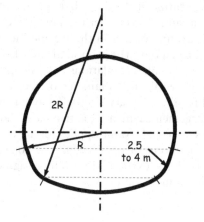

Figure 6.1 Example proportions of an SCL tunnel in soft ground.

As noted in Section 3.1.2, the lining tends to continue to deform until a complete structural ring is formed. Hence, early ring closure – i.e. within 0.5 to 1.0 tunnel diameters from the tunnel face – helps to reduce deformations and surface settlement. Typically, advance lengths are limited to around 1.0 m to facilitate early ring closure and stability at the face. The strength of numerical modelling is evident here as it permits different excavation sequences and geometries to be tested in the design phase.

As Figure 6.2 shows, the ratio of tunnel diameter to lining thickness generally lies between 10 to 15 for recent shallow SCL tunnels in the UK.

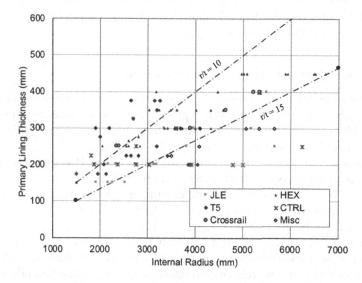

Figure 6.2 Lining thicknesses for SCL tunnels in soft ground.

There is considerable scatter in the data, but this is probably due more to differences in design assumptions rather than ground conditions.

The arching in the ground occurs in three dimensions around the active face. Hence load is thrown onto the ground ahead of the tunnel and backwards onto the lining. Simple calculations (e.g. Figure 4.2) indicate that arching occurs mainly within one diameter of the tunnel. Experience in the field supports this (e.g. Thomas 2003). To avoid interaction (e.g. in a pilot tunnel and enlargement sequence), active faces should be kept at least two tunnel diameters apart. The same applies for adjacent tunnels. If this is unavoidable (e.g. at junctions) or undesirable, the adjacent structures must be designed to cope with the extra loading. The design of junctions will be addressed in more detail in Section 6.5.

6.2 DESIGN FOR TUNNELS IN BLOCKY ROCK

This section covers the design of sprayed concrete linings in jointed rock masses (i.e. where the rock behaves as a discontinuum). The key mechanisms of behaviour are block failure, plastic yielding or general failure in the ground around the tunnel (see Table 3.3). The sprayed concrete may be needed for immediate support (e.g. for blocks or even as a full structural ring) but sometimes installation of the full sprayed concrete lining can be delayed for several advance lengths. Similarly, depending on the type of ground, the timing of ring closure varies, and in the best ground a structural invert is not required. Sprayed concrete is usually only one part of the support system, and it often functions in combination with rock bolts. As in soft ground, face stability, along with the available equipment, drives the choice of excavation sequence.

6.2.1 Key behaviour of sprayed concrete

The early age strength of sprayed concrete, especially its bond strength, is the key to supporting blocks. When pushed to its limit, under loading by a single block, a sprayed concrete lining usually fails first by debonding and then in flexure (Barrett & McCreath 1995) – see Figure 6.3.

Steel fibre reinforced sprayed concrete (SFRS) is very effective in supporting blocky rock masses. The fibres reinforce the full thickness of the concrete and absorb a lot of energy while deforming. SFRS is particularly suitable for high stress environments (e.g. Tyler & Clements 2004). Macrosynthetic fibres can also be used.

When acting as an arch or a full structural ring, the other aspects of behaviour, such as nonlinearity or creep, may be relevant (see Section 6.1.1). The importance of time-related characteristics such as creep depends on the rate of the development of the ground loads.

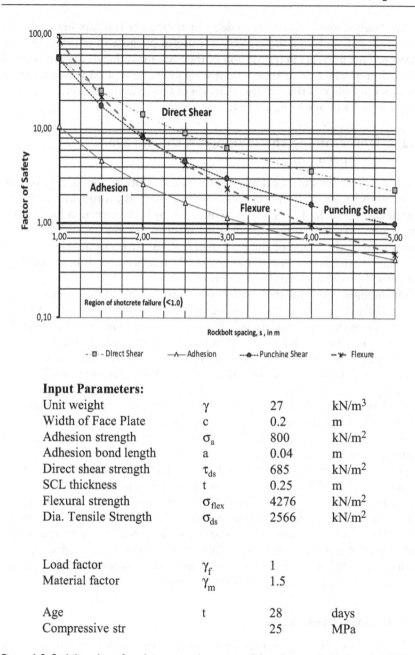

Figure 6.3 Stability chart for plain sprayed concrete (after Barrett & McCreath 1995).

Input Parameters:

Unit weight	γ	27	kN/m³
Width of Face Plate	c	0.2	m
Adhesion strength	σ_a	800	kN/m²
Adhesion bond length	a	0.04	m
Direct shear strength	τ_{ds}	685	kN/m²
SCL thickness	t	0.25	m
Flexural strength	σ_{flex}	4276	kN/m²
Dia. Tensile Strength	σ_{ds}	2566	kN/m²
Load factor	γ_f	1	
Material factor	γ_m	1.5	
Age	t	28	days
Compressive str		25	MPa

Although it may be possible to delay installing the full support, it should be noted that sprayed concrete serves a useful purpose in sealing the rock surface. This prevents deterioration of the rock mass due to drying out and the action of water or air on it, and it seals joints tight. This is as important for block stability as the structural action of the sprayed concrete, spanning between bolts or acting as an arch.

6.2.2 Determining the loading on the sprayed concrete

Table 4.2 contains some examples of analytical solutions for blocky ground. The disadvantage of semi-empirical methods such as Protodykianov's load distribution (Szechy 1973) is that, while they indicate a ground load or lining stress, they do not provide detailed information on the stresses in the lining. Given the shape of the load distributions, it is not a simple task to translate them into the distribution of axial forces and bending moments to be carried by the sprayed concrete lining.

Of the more sophisticated analytical solutions, the convergence confinement method (CCM), is the most commonly used in blocky rock. While it does assume that the rock mass behaves as a continuum, plasticity can be incorporated using the Hoek–Brown failure criteria which simulates rock better than other failure criteria such as Mohr–Coulomb. The CCM can be used as a quick check on the overall stability of the excavation and check for plastic yielding, bearing in mind its other limitations (such as the assumption of axisymmetry). It is often used to estimate convergence and experiment with the timing of installation of support.

Because of the better stability of the blocky rock, compared to soft ground, non-circular cross-sections can be used more often. This goes beyond the capabilities of most analytical solutions, so designers then turn to numerical modelling. The most well-known numerical modelling program for blocky rock is the discrete element program, UDEC, and its 3D companion, 3DEC. Some of the commercially available finite element or finite difference packages offer constitutive models that try to mimic the behaviour of jointed rock masses. Blocky rock presents an added complication for numerical simulation as it is difficult to determine many of the parameters that govern the behaviour of joints.

6.2.3 Lining design

In blocky rock, the support is designed either using empirical methods or by estimating loads by one of the means mentioned in the previous section, converting them into lining loads and sizing the sprayed concrete to carry them.

The most commonly used empirical methods are the Q-system and RMR (see also Sections 4.4.1 and 6.3.1). These work best in more competent rock, approaching hard rock conditions. At the lower end of the range, there is a risk that such simple methods will not provide a robustly engineered solution. Applying support designed for rock to failing soft ground is recipe for disaster. The borderline cases between blocky rock and soft ground deserve more detailed consideration. Also, it is worth noting that, like all empirical methods, they provide no information on the factor of safety or the timing of the support. The latter should be specified based on stand-up time.

If the sprayed concrete is acting as an arch or ring, it can be designed as a compression member under bending, in the same way as for soft ground tunnel linings. Depending on the strength of the rock mass and its jointing, a thin layer of sprayed concrete may be adequate to restrain small blocks, and it may be more appropriate to design the lining as a composite beam acting with the rock. The Voussoir arch theory makes good use of this composite action (Diederichs & Kaiser 1999, Oliveira & Paramaguru 2016). Bolts pin the reinforced sprayed concrete (which acts as a tensile membrane) to the rock beam. The axial load is transferred into "abutments" at the springing point of the arch. Amongst others, Asche & Bernard (2004) noted that this method works well in horizontally bedded competent rock provided that there is no sliding of blocks at the "abutments" (Banton et al. 2004) and there are no shallow dipping joints cutting across the horizontal bedding (Oliveira & Paramaguru 2016). Banton et al. (2004) describe the application of the Voussoir solution as well as other design tools for blocky rock.

If the sprayed concrete is only controlling the stability of individual blocks, effectively spraying between rock bolts, the analytical solution by Barrett and McCreath (1995)[1] can be used to size the lining thickness – see Figure 6.3. The blocks are held in place by adhesion over a strip around the perimeter of the block. This strip is assumed to be 30 mm wide, which enables blocks of 1.5 to 3.0 tonnes per linear metre to be carried (Banton et al. 2004). The computer program UNWEDGE offers a more comprehensive treatment of this aspect, since, based on inputted joint sets, it calculates all kinematically admissible blocks.

A tunnel usually passes through a variety of ground conditions. It is more economic to specify a range of support classes (typically three to six). The construction team can then choose the most appropriate one, depending on the actual conditions encountered.

The design of junctions will be addressed in more detail in Section 6.5.

6.3 DESIGN FOR TUNNELS IN HARD ROCK

This section covers the design of sprayed concrete linings in massive rock masses (i.e. where the rock behaves as a continuum). The key mechanisms of behaviour are usually a stable elastic response of the rock mass around the tunnel and isolated block failure (see Table 3.4). The sprayed concrete must provide immediate support to the blocks, but otherwise the tunnel is stable. As before, face stability, along with the available equipment, drives the choice of excavation sequence. Hard rock (in moderate to low stress environments) has long stand-up times so support may not be installed until several rounds behind the face. Round length is typically 3 to 4 m. The special cases of rockburst, swelling or squeezing rock are discussed separately below.

6.3.1 Lining design

Unless there are complications such as high *in-situ* stresses, empirical methods are often used to determine support (Grov 2011). These may be formal methods such as Q-system (see Figure 6.4) or RMR or simply the application of engineering judgement and informal "rules of thumb". It is preferable to assess systematically the risks posed by the ground conditions and use this in combination with the established support charts such as Q or RMR to determine the support. The timing of the support should be specified based on stand-up time. The excavation method too influences the support

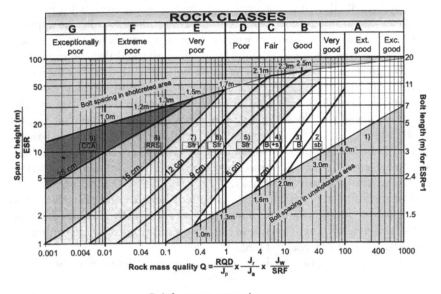

Reinforcement categories

1) Unsupported
2) Spot bolting
3) Systematic bolting
4) Systematic bolting with 40–100 mm unreinforced shotcrete
5) Fibre reinforced shotcrete, 50–90 mm, and bolting
6) Fibre reinforced shotcrete, 90–120 mm, and bolting
7) Fibre reinforced shotcrete, 120–150 mm, and bolting
8) Fibre reinforced shotcrete, >150 mm, with reinforced ribs of shotcrete and bolting
9) Cast concrete lining

Figure 6.4 Q-system support chart.

required. TBM tunnels tend to disturb the rock less than the drill and blast method and so require less support (e.g. Barton 2000).

The sprayed concrete, whether plain or reinforced with fibres or mesh, is usually only one component of the support system. Often it has a secondary role compared to the rock bolts. The capacity of sprayed concrete to carry blocks, either alone or in combination with rock bolts, can be checked by simple calculations (e.g. Barrett & McCreath 1995 – see Section 6.2.3) or programs like UNWEDGE. The CCM can be used as a quick check on the overall stability of the excavation, a check for plastic yielding and to estimate convergence.

For logistical reasons, small diameter rock tunnels often feature temporary enlargements for passing bays or mucking niches for temporary stock-piling of spoil. These simply extend the span of the excavation. They should be designed in the same way as the main tunnel but using the maximum span. The design of junctions will be addressed in more detail in Section 6.5.

As in all tunnelling, the *in-situ* stresses are redistributed around the tunnels. This may lead to overstressing of rock pillars between tunnels. Rock bolts are generally much more effective in reinforcing pillars than sprayed concrete. As expounded by Rabcewicz, the father of the NATM, and others, it is important to think of the support and the rock mass as one system which shares the loads. Rock bolts are more effective because they enhance the ability of the rock mass to carry the loads around the opening whereas the sprayed concrete functions more in supporting the surface of the tunnel.

For complex underground works or those with adjacent tunnels (e.g. powerhouses), it may be necessary to use numerical modelling in 2D or 3D.

6.4 SHAFTS

Linings for shafts can be designed in a similar manner to linings for tunnels. Intrinsically a shaft is more stable than a tunnel of the same diameter because gravity exacerbates stability problems in the crown of a tunnel. The same mechanisms of behaviour as outlined before for the three types of ground apply, except that more attention should be paid to invert stability. Sprayed concrete does not normally feature in the temporary measures to ensure the invert stability of the shaft. Subdivision of the excavation round is sometimes employed as is dewatering or construction of a pilot shaft (mined or by drilling and raise-boring). In loose water-bearing ground, vertical forepoling with steel sheets can also be used. Ground treatment or, in extreme cases such as deep mine shafts, ground freezing may be needed in the cases where there is a lot of water.

Sprayed concrete linings for shafts do not bear on the base of the shaft during construction but hang off the ground above using skin friction. Several analytical solutions exist for the determining the loads on shaft linings in soft ground. The basic CCM can be used although the solutions by Wong and Kaiser (1988, 1989) are more sophisticated. Notably they incorporate the effects of vertical arching at shallow depths around the shaft and horizontal arching at deeper levels. One phenomenon that is not included in their analytical solution is the reduction in loading at the base of shafts due to vertical arching that they observed in test results and measurements from real shafts. The standard methods described above can be used for support of individual blocks. In numerical modelling, an axisymmetric model can be used for single shafts. If there are junctions in the shaft or adjacent structures, it may be necessary to use a 3D numerical model. As you will have probably already guessed, the design of junctions will be addressed in more detail in Section 6.5.

Where sprayed concrete is used for the base of a completed shaft, the base is usually domed if a substantial water or ground pressure is expected, since this will help minimise bending in the sprayed concrete and transfer the load in compression into the lining of the shaft. One must also check the shaft for flotation.

Inclined tunnels are essentially a hybrid between tunnels and shafts in terms of design.

6.5 JUNCTIONS

Junctions – between tunnels and shafts or other tunnels – are an almost ubiquitous feature of tunnelling projects. One great advantage of SCL tunnelling is its efficiency in the formation of junctions. The SCL structure acts as a shell, transferring the stresses around the opening. This is believed to be aided by favourable aspects of the behaviour of sprayed concrete such as creep. By a combination of this stress redistribution and the promotion of arching in the ground, a well-designed SCL junction avoids excessive stress concentrations at the edge of opening. However, there will be some increase in stresses so the area around an opening and at the start of the new tunnel may need to be strengthened to cope with this.

This section should be read in conjunction with the relevant section for the type of ground that the junction is located in. The same basic principles apply with some additional ones as outlined below.

6.5.1 Key behaviour of sprayed concrete

As noted, depending on the ground, different aspects of the behaviour of sprayed concrete will assume differing degrees of importance. In addition to this, assuming that the sprayed concrete lining is acting as a structural ring, at a junction the age-dependent behaviour of the sprayed concrete in the new ("child") tunnel and creep in both linings should be considered. These phenomena produce a "softer" response, resulting in more deformation and stress redistribution than a stiffer material would. Provided that the overall stability of the junction is maintained, this produces effective stress transfer around the opening. Depending on how old the existing ("parent") tunnel is, age-dependent behaviour in its sprayed concrete may or may not be relevant.

6.5.2 Determining the loading on the sprayed concrete

The key for the designer is predicting how the stresses in the lining and the ground will be redistributed as the new tunnel is built. This process is not well understood, and so junctions can appear difficult to design. Junctions increase the effective span of the excavation. Consequently, the loads on the lining around a junction are higher than normal.

The stronger the ground is, the less work the sprayed concrete lining will have to do. In very competent, hard rock, rock bolts alone may be used to reinforce the ground and secure the junction. If used at all, sprayed concrete will merely serve to secure individual blocks on the surface of the excavation. In weaker ground, the sprayed concrete lining carries more of the ground load and must function as a full structural ring.

Simplistic design tools – such as beam-spring models – are not suitable for simulating the construction of junctions as they are incapable of replicating realistically the stress redistribution in the ground.

6.5.3 General arrangement and construction sequence

Understanding the three-dimensional arrangement of the junction is critical. Due to arching in the ground and the lining, the junction will interact with any nearby objects, i.e. within about one tunnel diameter. One effect of the junction is to increase the effective span of the opening in the ground, which should be considered when using 2D methods in the design. Unfortunately, there is little published guidance on the design of junctions (e.g. on the relative size of the two tunnels or proximity between adjacent junctions). There is a recommendation in the Q-system to multiply the joint number by three at junctions (Barton et al. 1975). On the other hand, there are many examples of the successful use of SCL tunnel junctions.

Often there are several phases in the construction of a junction. In preparation, it is good practice to form a tunnel "eye". The eye is the area which will be broken out. To make this easier, it is usually thinner and more lightly reinforced than the rest of the parent tunnel. If possible, starter bars can be added to provide continuity of reinforcement when the new tunnel is started. The area around the eye is then strengthened. This can be done in several ways (see Figure 6.5):

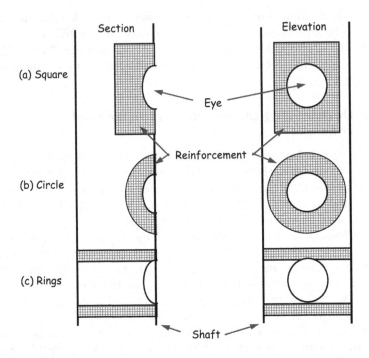

Figure 6.5 Reinforcement around junctions.

- Reinforcing a patch around the eye. A square patch is easier to construct than a circular one, not least because the reinforcement can be added as pieces of mesh as the parent tunnel advances. For circular patches, bars are added later individually in a circumferential/radial configuration or as rectangular pieces of mesh.
- Adding reinforcing rings to either side of the opening

The latter is both less elegant and less effective in transferring stresses smoothly around the opening. The rings tend to be large sections so they may intrude on the internal space of the parent tunnel or the tunnel itself has to be enlarged locally.

Several construction sequences are commonly used (see Figure 6.6):

- Stub
- Pilot tunnel and enlargement
- Reinforcement of the "child" tunnel
- The normal excavation sequence for the tunnel

The first two are used in weaker ground where the priority is to ease the birth of the child tunnel. Subdivision of the tunnel face improves stability, and the early installation of support reduces ground movements. The

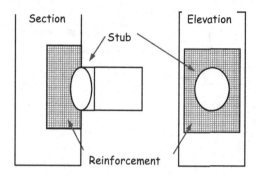

Figure 6.6 Construction sequences for junctions.

formation of a full structural ring (e.g. in the form of a stub) should arrest the deformation of the tunnel linings. Having stabilised the junction, the new tunnel can then progress as normal. Alternatively, the first part of the new tunnel (a distance equivalent to half a diameter) can be thickened and/ or reinforced to cope with the extra loads. The final sequence above should only be used where ring closure can be achieved in the child tunnel quickly since it is supporting the parent too.

6.5.4 Lining design

In the past, the sprayed concrete lining at a junction has been designed on the basis of precedent or simple models such as the "hole in an elastic plate" analytical solution (see Figure 6.7).

Knowing the loads in the lining of the parent tunnel, the latter can used to model the redistribution in the sprayed concrete shell. The effect of adjacent junctions can be estimated by simply assuming superposition. The obvious limitations are that the plate is flat, unlike the curved lining, and it is assumed that the total ground load remains the same. Another important question when using this method is: what is the stress in the longitudinal direction? Any stress in this direction will reduce the peak tangential stresses at the edge of the opening although the shear stresses increase (see the effect of varying K in Figure 6.7).

These new loads are used to determine the additional reinforcement around the opening. Rather than take the highest forces, which occur at the edge of the opening, sometimes the forces are averaged over the area to be reinforced. Typically, this is 1 to 2 metres from the edge of the opening. It can be argued that creep and nonlinearity in the sprayed concrete will tend to smooth out stress concentrations. Bending moments can be derived by assuming the axial forces act at a nominal eccentricity (typically 20 mm – Eurocode 2 (2004)[2]). Overall this method is viewed as being conservative as it tends to predict thicker linings and heavier reinforcement than precedent

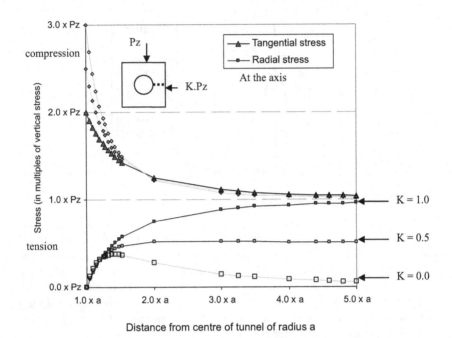

Figure 6.7 Stress distribution around a hole in an elastic medium under applied stresses Pz and K.Pz.

practice suggests is necessary. One study of monitoring data from a junction has also supported this view (De Battista et al. 2015). Often the lining of the "child" tunnel is not increased beyond its normal thickness, on the basis than all additional loading at the junction is carried by the (larger diameter) "parent" tunnel.

A minor refinement of the method above is to use a 2D numerical model of the opening in the flat plate. As well as being able to include non-circular geometries, this opens the door for nonlinearity and creep to be included in the model.

The final option is the use a 3D numerical model (see Figure 6.8). Before embarking on this, it is worth remembering that this will prolong the design process. The geometry of a junction is more complex so it takes longer to build the model and requires more elements. To get the best out of the model, the constitutive modelling should be as realistic as possible. As a result, the models tend to be large and therefore slow to run. Interpretation is made more complicated by the volume of data available and the fact that the pattern of stress distribution tends to change as the child tunnel progresses away from the parent tunnel.

However, with modern computing power, 3D models are seen more and more often in design. As an example, on one recent major project, a

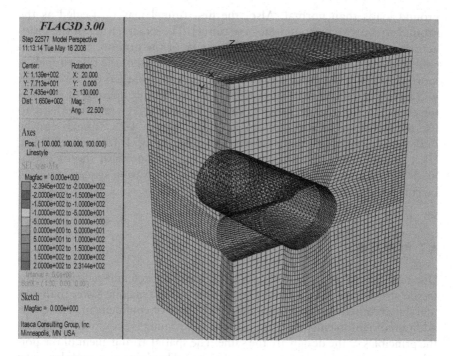

Figure 6.8 3D numerical model of a tunnel junction.

parametric study was run with 3D numerical models, and from the results a simple design chart was successfully produced to estimate the additional strengthening needed around the junctions. The "hole in an elastic plate" solution was found to agree well with the sophisticated model in terms of the concentration of hoop force and moment at the axis in the parent tunnel. However, the picture above and below the opening in the hoop direction was less consistent with that solution or experience on site. The child tunnel appeared to drag the lining of the parent into tension, which is an effect seen by other researchers (e.g. Jones 2007).

As with all other structures, care should be taken with the detailing to keep the construction as simple as possible. For example, concentrations of reinforcing bars should be avoided.

The potential for differential movement at junctions between SCL tunnels and other structures should be considered. Special measures may be required to avoid structural damage or water ingress.

6.6 TUNNELS IN CLOSE PROXIMITY

Building a tunnel modifies the stress in the ground around it. Therefore, if another tunnel is built close to the first one, it may encounter higher ground loads than normal. Also, the first tunnel may be loaded as the ground stresses arch around the new tunnel. The exact nature of the interaction depends on the stiffness and strength of the ground. Generally, in soft ground, if the separation between the centrelines of the two tunnels is more than twice the largest diameter of the tunnels, then the interaction will have a minor effect. Figure 6.7 shows one simple method of estimating the impact of the interaction, although it is arguably more realistic to use the plasticity solution for this case (see Figure 4.2). Tunnels can be built very close to each other even in soft ground (e.g. JLE London Bridge, UK), provided that the linings are designed to cope with the increased loads. When tunnelling in close proximity the excavation sequence should be considered carefully.

6.7 PORTALS

Due to the low cover, the ability of the ground to arch and redistribute the *in-situ* stresses is limited. The strength of the ground may be lower than normal due weathering, slope instability or the lack of confinement. Therefore, additional reinforcement of the ground and/or lining is required. Otherwise the principles of the design are the same. The spatial arrangement of the portal and support measures must be considered. The lining at a portal may be subjected to pronounced asymmetric loading if the tunnel intersects with a slope or rock head obliquely. There is a recommendation in the Q-system to multiply the joint number by two at portals (Barton et al. 1975). Since they are important structures, portals may warrant the use of numerical modelling, ideally in 3D.

It is prudent to specify more conservative excavation sequences, firstly to prevent instability but also because the portal is built at the start of tunnelling while the construction team is still in the learning phase. Additional support measures such as steel arches, spiling or canopy tubes are commonly used at portals.

6.8 SPECIAL CASES

When confronted with a special case in design a common response is first to scour the published literature to find similar case histories and ideally guidance or rules of thumb. Combined with the basic principles, this information can be used to identify an outline design solution, which is then further developed with the help of numerical modelling. In the sections below some special cases are considered briefly. The comments are not meant to be comprehensive but rather cover only the impact on the design of the sprayed concrete lining.

6.8.1 Seismic design

Tunnels generally perform well in earthquakes. They are flexible enough to move with the ground and, being embedded in the ground, they are much less vulnerable to the effects of their own inertia, unlike structures above ground. The risk of damage decreases as the height of overburden increases (Hashash et al. 2001). Although the tunnel may not suffer major structural damage, the contents of the tunnel might not fare so well. The worst cases for a tunnel are when it intersects a fault or where there are substantial changes of stiffness in the structure (e.g. at tunnel junctions and portals). In both cases the lining and/or the waterproofing could sustain major damage. A detailed review of the effects of seismic events on tunnels in general can be found in Hashash et al. (2001).

In short, an SCL tunnel is likely to be subjected to three types of displacement:

- Axial compression and extension – due to seismic waves running parallel to the tunnel
- Longitudinal bending – due to seismic waves which move the ground in a direction perpendicular to the tunnel axis
- Racking/ovalisation – due to shear waves normal to the tunnel

A simple solution to the intersection with a fault is to build an enlargement at the intersection (e.g. Los Angeles Metro and San Francisco BART metro in the US). If there is movement at the fault, there will still be enough space for tunnel to remain operational, although the internal works will require adjustments, e.g. re-aligning the track in a rail tunnel. A similar approach was adopted on the Channel Tunnel at Castle Hill (Penny et al. 1991). The tunnel diameter was increased by 1.22 m, to allow for possible movement along the line of a landslip. Movement joints were installed and the lining reinforced near the failure plane.

Where there is a large difference in stiffness, there is potential for damaging differential movement. Movement joints are installed to accommodate this (e.g. see Figure 6.9). Standard texts on seismic design should be consulted for more information on such movement joints.

As for the lining, Hendron and Fernandez (1983) proposed a simple check to determine whether or not the tunnel is substantially stiffer than the ground in terms of its flexibility, F (see Equation 6.1).

$$F = \frac{2E_g \cdot (1-v)R}{E(1+v_g)t} \tag{6.1}$$

where E_g and v_g are the elastic modulus and Poisson's ratio of the ground, E and v are the elastic modulus and Poisson's ratio of the lining, t is the lining thickness and R is the radius of the tunnel. If F is greater than 20, the lining

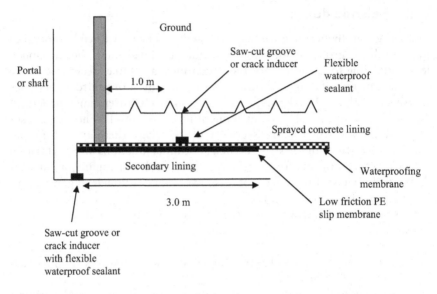

Figure 6.9 Movement joint.

can be deemed to be perfectly flexible compared to the ground and it should not experience damage.

A more detailed analytical approach, originally proposed by Wang, is reproduced as a worked example in Hashash et al. (2001). For more detailed analyses, there are commercial numerical modelling programs, which can simulate dynamic behaviour. The effects of dynamic loading on sprayed concrete are discussed in Sections 3.3.2 and 6.8.5.

6.8.2 Squeezing ground

Squeezing ground (like swelling, creep and rockburst) essentially presents an additional form of loading. Hoek and Marinos (2000) described squeezing in detail and proposed an equation of estimating squeezing potential (see Equation 6.2). They suggest that, in unlined tunnels, squeezing begins to pose problems when the *in-situ* stress exceeds 2.25 times the rock mass strength. At stresses above 3.5 times the rock mass strength, severe squeezing may occur, with the convergence of the tunnel lining potentially exceeding 2.5% of the diameter, i.e. 250 mm convergence in a 10 m diameter tunnel. This represents huge forces which can easily destroy a stiff lining that is placed in the tunnel. Indeed, it is just these conditions that led Rabcewicz to formulate the NATM approach.

$$\varepsilon = 100 \cdot \left(0.002 - 0.0025 \frac{p_i}{p_o} \right) \cdot \frac{\sigma_{cm}}{p_o}^{\left(2.4 \frac{p_i}{p_o} - 2 \right)} \qquad (6.2)$$

where ε is maximum strain in the ground (defined as tunnel convergence/tunnel diameter), p_i is internal pressure (resistance) provided by the lining, p_0 is *in-situ* stress and σ_{cm} is the uniaxial compressive strength of the rock.

Rather than trying to resist the forces of nature, Rabcewicz advocated installing a lining that was flexible enough to absorb these large deformations and would bring them to a halt in a controlled manner. Sprayed concrete with its pronounced time-dependent behaviour works well in these situations when used in conjunction with rock bolting which reinforces the rock mass around the tunnel. Installation of the secondary lining is delayed until the deformations have stabilised. However, in the worst cases, this may not be enough, and slots are left in the lining to permit the large convergences to occur without wholesale destruction of the sprayed concrete lining. As described earlier (see Section 3.2.3), the slots may be left empty or yielding supports can be inserted to help control the deformation. Slots or yielding supports can be modelled in numerical simulations by gaps in the lining or elements with a low stiffness (e.g. see Arlberg tunnel in John (1978) and the fault zone in the Sedrun section on the Gotthard Base Tunnel (Henke & Fabbri 2004)).

6.8.3 Swelling ground

Swelling is associated with marls, anhydrite, certain basalts and clay minerals such as corrensite and montmorillonite. Swelling is a stress-dependent process. It can be minimised by limiting the exposure of the ground to water and maintaining confinement. For example, the swelling of anhydrite is caused by the absorption of water, and the stress relief around a tunnel causes fissures to open and the permeability of the rock mass to increase, thereby letting water into the anhydrite. The Huder–Amberg test can be used to determine the stress–strain relationship for the ground, and this can be used to estimate the swelling loads on the lining. However, it is best to validate these estimates with measurements from a real tunnel as the behaviour may differ. For example, there is some evidence that the swelling may cause "self-healing" as it reduces the permeability of the rock mass and so the penetration of water. The website of Prof W. Wittke (www.wbionline.de) is a useful source of information and references. Examples of tunnels in swelling ground include the Lyon–Turin Base Tunnel (Triclot et al. 2007) and the Freudenstein experimental gallery (www.wbionline.de).

As in the case of squeezing ground, here the installation of the secondary lining is often delayed until the deformations have stabilised. Convergences can be as large as 1 or 2 m, with the result that the tunnel has to be reprofiled (Triclot 2007). Even so a thick and heavily reinforced secondary lining may be required to resist the residual swelling pressures. Alternatively, in more extreme cases, a yielding support may be installed. This can include yielding arches in the primary lining (Triclot 2007) or a compressible invert of the secondary lining (Wittke 2007), in addition to deep rock bolt reinforcement. A ring of face dowels reinforcing the ground ahead has also been found to be beneficial (Triclot et al. 2007).

6.8.4 Creeping ground

Rocks such as rock salt, chalk, coal and marl may exhibit creep with the result that over time the ground will continue to deform and add load to the tunnel lining. The extent of creep in the ground is also heavily dependent on the stresses in the ground.

Such behaviour should be considered in the design calculations. Many commercial numerical modelling programs offer constitutive models for creep behaviour. As with all advanced numerical modelling, where possible the components of the model should be validated. For example, the suitability of the creep model of the rock could be checked by back-analysing laboratory creep tests or field data and comparing the predictions with the test results (e.g. Watson et al. 1999). The creep of the ground will interact with the creep of the sprayed concrete lining. Alkhaimi (1995) noted that in this case the creep of the sprayed concrete may exacerbate the situation rather than being beneficial. Hellmich et al. (1999c) found that the loads on the lining will increase as the rate of creep in the ground decreases relative to the rate of hydration, as more load is being added to the lining at later times when it is stiffer.

As in the case of squeezing ground, here the installation of the secondary lining is delayed until the deformations have stabilised. However, the creep may continue over a long period so a thick and heavily reinforced secondary lining may be required to resist the residual pressures. The Channel Tunnel Crossover is an example of an SCL tunnel built in creeping ground (Hawley & Pöttler 1991).

6.8.5 Rockburst

Rockburst is the sudden failure of rock due to overstressing. It tends to occur soon after excavation, but it may continue over a long period. Rockburst can be very dangerous not just because of the violence of the spalling but also because it is very unpredictable. Grimstad (1999) suggests that rockburst is worst in stiff, strong rocks which tend to behave more brittlely than weaker ones. Several texts contain detailed descriptions of the phenomenon and countermeasures (e.g. Hoek & Brown 1980, Kaiser & Cai 2012). Kaiser and Cai (2012) list the three functions for rock support in the case of rockburst: reinforcing the rock mass; retaining broken rock and tying the fractured blocks back to stable ground. Given the complexity of the phenomenon, it is difficult to offer simple recommendations. Kaiser and Cai (2012) eloquently describe the design process for these cases. Rockburst imposes a dynamic load on the support. So-called "dynamic" bolts may be needed instead of normal ones. The sprayed concrete may be damaged due to dynamic effects as seen in blasting (see Section 3.3.2) so alone it is unlikely to be sufficient as support. '

The Laerdal Tunnel in Norway is one tunnel which experienced rock-burst during tunnelling (Grimstad 1999), and there it was found best to install the rock bolts after first spraying the concrete. Both steel fibre and synthetic fibre reinforced sprayed concrete, in combination with rock bolts, have been found to perform well in this extreme situation (Bernard et al. 2014) because FRS has a high capacity for absorbing energy during deformation and because it can withstand larger deformations than standard steel mesh either on its own or even embedded in sprayed concrete (see ITA 2006 (report from South Africa) for further information). Other studies have suggested that sprayed concrete with mesh reinforcement (especially high tensile strength mesh) (ITA 2010) or TSLs may perform better than using fibre reinforcement alone (e.g. Archibald & Dirige 2006).

6.8.6 Compensation grouting

When tunnelling in sensitive urban environments, it is increasingly common to see compensation grouting used to mitigate the effects of tunnelling-induced ground movement. Compensation grouting involves pumping cement grout at high pressure into the ground to force the ground upwards and reverse the settlement. When applied near tunnels, this can impose considerable loads on tunnel linings, and it has even been cited as a contributory factor in at least one tunnel collapse (HSE 2000). The increase in loads is particularly high when grouting is performed at less than half a diameter from the tunnel, but the magnitude of the impact depends on a number of other factors such as the geology and grouting pressure.

Either the lining can be designed to cope with the additional loads (in this temporary case) or a system of exclusion zones can be imposed to protect the most vulnerable part of the tunnel (the active face with its incomplete ring) from the loads or a combination of both steps can be taken.

As an example, compensation was used extensively on the Crossrail project in London. The design team performed a series of numerical models to explore the impact of compensation grouting on lining loads and deformations. The sensitivity study was calibrated against some data from previous projects in London. A design chart was developed to assess the impact on loads and to determine trigger levels for deformations. Typical guidance in this case recommended that no grouting be performed near an open face (see Figure 6.10). While this work remains unpublished at present, it clearly demonstrated the power of numerical models in creating simple design methods for complex cases, and the project was completed successfully.

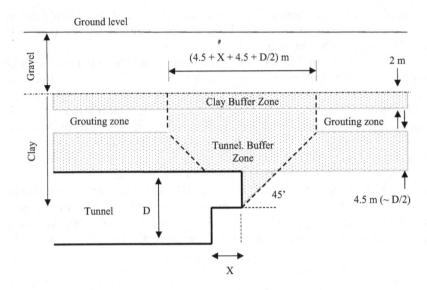

Figure 6.10 An example of exclusion zones for compensation grouting.

6.8.7 Compressed air tunnelling

Occasionally sprayed concrete has been used in the construction of tunnels under compressed air (e.g. Munich & Vienna Metros in 1980s – Strobl 1991). In itself, this construction method does not really affect the design of the sprayed concrete lining. In practical terms, air loss through the spray concrete is a concern for the construction team. Compressed air is usually employed in water-bearing granular material so the more general questions of the suitability of sprayed concrete in this case also apply. The compressed air will keep the groundwater out but the effectiveness of the lining will depend on how easy it is to spray onto the ground.

As a whole, there may be an increased risk to the tunnel as the system is not "fail-safe". If there is a massive loss of air, for example, through an area of loose ground, the pressure will drop and water will flow in, further weakening the ground. Spraying concrete on the face may not be able to stabilise the ravelling ground.

Loss of the air through the ground and the lining increases the costs of compressed air tunnelling. This can be reduced by treating the ground (e.g. permeation grouting or spraying a sealing coat of concrete on the face) or improving the quality of the sprayed concrete (e.g. by adding microsilica, increasing early age strength, curing and reducing the water–cement ratio) (Bertsch 1992).

The air inside a tunnel under compressed air tends to be quite humid which can pose problems for dry mix sprayed concrete as it may start to hydrate before it is sprayed (ÖBV 1998).

6.8.8 Frozen ground and cold weather

A cold environment slows the hydration but the final strength is not usually reduced by much so long as the concrete does not freeze while hydrating. One simple remedy is to heat the tunnel and the rock surface so that the concrete gains strength sufficiently quickly. As an example, in Norway, the temperature of the rock should be greater than +2°C and the temperature of the concrete mix itself more than +20°C when spraying (NCA 2011). Figure 6.11 shows the use of heaters and plastic sheeting during curing. The temperature of sprayed concrete should not be allowed to drop close to freezing during curing as this could stall the hydration process and the formation of ice would disrupt the structure of the young concrete.[3] Once the concrete has reached a strength of about 5 MPa such protective measures can be stopped. ACI 506R-16 (2016) sets a lower threshold of 3.4 MPa.

Sprayed concrete is not often exposed to cold temperatures in a tunnel unless ground freezing is being used to stabilise the ground. In that case the early age strength of the concrete is less important structurally as the freezing helps to support the ground, but the concrete is accelerated to speed up hydration and to reduce the risk of freezing during this sensitive period. Sometimes it is necessary to pre-heat the ingredients or to use hot water too. Klados (2002) provides a detailed description of one case study, including the mix design and profiles of temperature in the lining.

In colder climates, the tunnel lining within about 200 to 400 m of the portals can be exposed to freeze–thaw cycles (ITA 1993) – depending on

Figure 6.11 Weather proofing in icy conditions.

the climate and type of tunnel. Spraying can entrap air as well as entraining it, and the compaction on impact forces some of that air out. Provided that there are sufficient pores of the right size in the concrete, the freezing of water in the concrete will not cause damage. ACI 506R-16 (2016) recommends that the water–cement ratio is less than 0.45 for the mix, the entrained air content in place is greater than 3% and the maximum air-void spacing factor is 0.254 mm. To achieve this, normally the air content before pumping must be more than 6%.

6.8.9 Hot ground and hot weather

The temperature of the ground increases with depth, at a rate of about 20°C per 1,000 m. So, in extreme cases, the rock itself can be hot. For example, rock temperatures of up to 45°C were encountered on the Gotthard Base Tunnel project (Greeman 2000). This can reduce the strength of the sprayed concrete. An equation for predicting the reduction due to high temperatures is given in Section 2.2.1. Although the temperature in linings in shallow tunnels rises to more than 40°C during hydration, this rise is short-lived. This is less onerous than the prolonged curing at elevated temperatures in deep tunnels. Therefore, thermal damage in SCL tunnels is probably negligible, except in the extreme cases of very hot ground.

Obviously, hot weather is unlikely to affect tunnelling works, except for exposed areas of the portal. In hot climes, the standard procedures should be adopted to avoid damage to the concrete due to excessive shrinkage. Care should be taken during the batching and delivery of the concrete to ensure that it does not start to hydrate before it reaches the tunnel. Finally, in hot countries, the normally available cements may have a reduced proportion of tricalcium aluminate or a lower Blaine value to slow the hydration process for outdoor applications. This would result in a slow setting sprayed concrete and could cause problems with low early strengths. However, the normal pre-construction testing should identify any potential problems (see Section 7.2.1).

6.8.10 Fire resistance

The subject of fire resistance of tunnel linings is an extensive and growing subject, spurred on by the consequences of some major fires in tunnels in the past. It is also intertwined with the fire-life-safety strategy for the tunnel operation as a whole. Guidance on the subject has been produced by the ITA (2004, 2017), albeit only for road tunnels at this time. Only brief comments are provided here by way of introduction to the subject. A similar approach applies to other types of tunnels.

The principles behind design for fire resistance demand, in the first instance, that the tunnel lining can withstand a fire for a certain time so that the people in the tunnel can be evacuated, typically 60 to 120 minutes.

In addition, if the tunnel passes under high risk structures such as a railway line or is located under the sea/a body of water or in unstable ground, where the consequences of a lining collapse are unacceptable, then the lining must also be able to survive the fire without major damage. The author is not aware of any example of a fire in a public tunnel that has caused the tunnel to collapse, although there are examples of fires in mines which have caused local collapses. The ITA guidelines propose a range of design criteria based on the type of tunnel and the traffic within it. For the design analyses, sprayed concrete can be assumed to behave in the same way as normal concrete. Winterberg and Dietze (2004) describe the mechanisms that cause damage to linings during fires, and they reviewed the assumed fire curves that can be used for design.

Quite often, where sprayed concrete forms the permanent lining, fine polypropylene fibres have been added (typically 1 to 2 kg/m^3) to enhance the fire resistance (e.g. Heathrow Terminal 5 – Hilar & Thomas 2005). In the event of a fire, the fibres melt and let the steam which is generated inside the lining by the heat of the fire to escape without causing explosive spalling (ITA 2004, Winterberg & Dietze 2004). This method was first used for tunnel lining segments and has been found to perform well in fire tests (ITAtech 2016) – see also Section 2.2.8. Coarse polypropylene fibres or structural synthetic fibres do not exhibit this beneficial effect.

6.9 SPECIFICATIONS

Specification of sprayed concrete works is relatively straightforward since there are several published guides (see Table 6.1) that can be used as a basis for a project's specification. A good specification should be comprehensive yet concise and unambiguous. Typically, it defines the inputs required (i.e. materials and competences of key staff), methods (both how to build the structure and the management processes) and the quality of the final product (e.g. strengths, geometric tolerances or watertightness). Because sprayed concrete linings are formed *in-situ*, not in the controlled environment of a factory, there tends to be more detail in a specification. The form and detail

Table 6.1 Common specifications

Specifications	Country
EN 14487 (2006) and EN 14488 (2006) along with Eurocodes	EU
ACI 506.2 (2013) and ACI 506R-16 (2016)	USA
British Tunnelling Society (BTS) Tunnel Specification (2000)	UK
EFNARC (2002)	Europe
Österreichischer Beton Verein (2013)	Austria
Norwegian Concrete Association (2011)	Norway

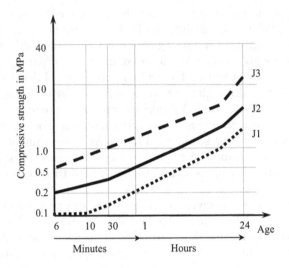

Figure 6.12 Early age strength criteria.

of a specification is governed by the unique needs of each project. However, some general comments have been included below.

Target strengths should always be specified for the early age period (i.e. t < 24 hours), for example, using the J-curves in EN 14487-1 (originally published by ÖBV (1998) – see Figure 6.12). In some countries, there are specific strength thresholds to be met before workers can enter the area under freshly sprayed concrete (e.g. 0.5 MPa on Crossrail (King et al. 2016), 1.0 MPa in Australian mines (Gibson & Bernard 2011, Rispin et al. 2017)) or a system of exclusion zones is used (CLRL 2014). The J2 curve is the minimum criteria for structural sprayed concrete. J1 refers to concrete that has no special load-bearing requirements during the first 24 hours, while J3 refers to special cases where a very rapid set is needed (e.g. to control strong water ingress). It is vital that the sprayed concrete gains sufficient strength to carry the anticipated loads at all ages.

Fibre reinforcement is addressed in detail in EN 14487 as well as the guidelines from ACI 506.2-13 (2013), ÖBV (2013) and EFNARC (1996).

In addition to the engineers and foremen, the key staff for an SCL tunnel are the nozzlemen (and, albeit to a much lesser extent, the pump operator) and requirements for their competencies should be specified (e.g. see ACI 506R-16 2016). Several programmes exist for nozzleman certification, most notably the EFNARC scheme for wet mix robotic spraying (Lehto & Harbron 2011, www.efnarc.org). A review of existing schemes can be found in Larive and Gremillon (2007). The competence of the engineering staff is also important since the effectiveness of decision making (either at the tunnel face or in daily review meetings) is governed by their judgements. Many of the failures of SCL tunnels stem in part from the inexperience or poor judgements of engineering staff. For SCL tunnels, it is important that the site team understands the methods in use, the limits of the design and how the tunnel is intended to behave. Setting criteria for the competence of the site team goes some way to achieving this, but there must also be good communication between the designers and the site team. This can be aided greatly by having a design representative on site (see Section 7.4).

The quality control test regime stipulated in the specification should be appropriate to the scale of the works (see Section 7.2), especially on small projects (Kavanagh & Haig 2012). Specifications can be made more user-friendly by adding in tables to list the required tests and their frequency as well as a table of submittals and their timing (e.g. EN 14487-1 2005).

Furthermore, the list of tests should be consistent and aligned with the design. One should not use a "pick-and-mix" approach of selecting fragments from different specifications since they may result in an incoherent testing regime. As a note of caution, some American specifications are geared more towards surface applications than tunnelling.

6.10 DETAILING

6.10.1 Steel reinforcement

To reduce the risk of shadowing, the minimum spacing between bars is typically 100 mm. Bars should be placed close to but not closer than 10 mm to the substrate so that the concrete can be sprayed behind the bars. Some guides put limits on the maximum diameters of bars to be used (e.g. 14 mm (ÖBV 1998), 12 mm (Holmgren 2004)) while Fischer and Hofmann (2015) suggested 16 mm (see also Section 3.4.4). Bars up to 40 mm have been sprayed in successfully, but this is very difficult to do, especially where the bars lap or cross. Also, the thicker the bars, the harder they are to place in the tunnel because they are heavier and more difficult to bend.

6.10.2 Structural continuity at joints

The normal rules for structural continuity of reinforcement apply and lap lengths can be calculated from standards for cast *in-situ* concrete. Often more informal rules are used. For example, the overlap of mesh reinforcement is normally expressed as a multiply of "squares" of the mesh – typically two squares. This is an easy rule of thumb for miners to remember.

At some joints (e.g. the footings of headings) it is not possible to lap both layers of mesh directly. Traditionally L-shaped bars are used at joints to provide an overlap (see Figure 6.13). When the new section of the tunnel is excavated, the L-bars are exposed and bent straight. The disadvantage of this system is that the placement of the individual bars is time-consuming and it can be hard to locate them later. To overcome this, prefabricated

Figure 6.13 Connecting reinforcing bars at a footing joint.

strips can be used – e.g. KWIK-A-STRIP – see Figure 6.14. They are quick and simple to use so the quality is better at the joints.

The bond of the concrete too is important at joints for structural continuity. The geometry of the joints should be designed so that they are easy to spray but also do not introduce a potential plane of weakness into the lining. On one major project, the ambitious target was set that the joints in

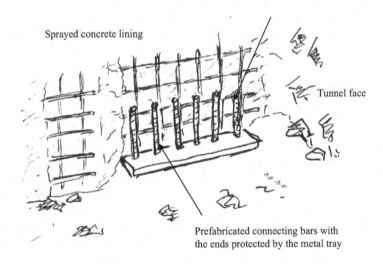

Figure 6.14 KWIK-A-STRIP at a joint in the lining.

the steel fibre sprayed concrete lining should have the same tensile capacity as any other part of the lining. Tests proved that this was achievable (Hilar et al. 2005). Similarly, Trottier et al. (2002) found that construction joints did not detrimentally affect the flexural capacity of test panels of plain, synthetic- or steel-fibre reinforced sprayed concrete. In contrast, the initial load capacity (i.e. load at first crack) was lower for mesh reinforced panels with joints than unjointed panels.

6.10.3 Waterproofing at joints

Waterproofing is a huge subject in its own right. It is discussed in part in Section 4.2.3. This section covers only waterproofing details for joints in sprayed concrete.

Reinjectable grout tubes have become popular in recent years as a joint sealing measure at major joints (such as at junctions – see Figure 4.12). A big advantage is that they can be used repeatedly, provided that the tubes are flushed clean after each injection of grout. This is useful where differential movement may occur over a prolonged period of time.

Figure 6.15 Drip-shed inside an SCL tunnel.

Hydrophilic strips are also sometimes used at major joints. In theory, since they are confined within the concrete, when water meets them they will swell and seal the water path. The strips must be kept dry until covered in concrete to prevent premature expansion.

Both of the above may be used with crack-inducers to improve their effectiveness. The traditional water-bars used for cast *in-situ* concrete are not suitable for sprayed concrete because it is too difficult to spray the concrete around them.

In some cases, linings have been designed to rely solely on the watertightness of the sprayed concrete (e.g. tunnels at Heathrow Terminal 5 – see Hilar & Thomas 2005). This is only suitable when the ground itself is largely impermeable (see Section 4.3.4). It is difficult to ensure that there is no cracking or that a good bond it always achieved at the many joints in a sprayed concrete lining. However, any residual leakage can be countered with conventional injection techniques for sealing leaks in concrete. Alternatively, a drip-shed can be installed inside the tunnel to prevent water appearing in the operational part of the tunnel. This is a common approach in Scandinavian tunnels (see Figure 6.15).

NOTES

1. NB: there appears to be an error in the equation for flexural capacity. It is suggested that, more correctly, $C_{flex} = \sigma_{flex}.(t^2.s)/6$ – see original paper for the definition of the terms in this equation (see also Bertuzzi & Pells 2002).
2. Eurocode 2 (2004) Cl. 6.1(4) sets the eccentricity as 0.033 × the lining thickness but not less than 20 mm. Linings are less than 600 mm so in practice this means 20 mm.
3. There is some limited evidence that accelerated sprayed concrete is less vulnerable to damage by freezing because of the way that the water is bound inside the ettringite (Beck & Brusletto 2018).

Chapter 7

Construction management

Because a sprayed concrete lining is manufactured in the tunnel and because of the opportunity to vary elements of the design, rigorous management of construction assumes an even more important role than in other types of tunnels, such as a segmentally lined TBM tunnel, where the segments are manufactured in a factory and installed within the safety of a TBM shield. Construction management includes the control of quality and monitoring to verify that the SCL tunnel is behaving as the designer had intended. These topics will be discussed below. As noted earlier (see Section 6.9), the competence of staff has a major influence on quality and minimum requirements should be specified.

7.1 SAFETY, OCCUPATIONAL HEALTH AND ENVIRONMENT

The use of sprayed concrete introduces a number of new hazards and influences some pre-existing hazards in tunnelling. The same methods used to assess safety can be used for SCL tunnelling. Risk assessments are the best tool for evaluating the safety of all aspects of tunnelling – not just in the construction phase but throughout the tunnel's life (see also Section 4.1.1). In this section, some of the major hazards will be highlighted. Anyone involved in a real tunnel should ensure that they understand how the relevant local safety regulations pertain to their specific project.

7.1.1 SCL tunnelling in general

The main risks relate to the stability of the tunnel in the prevailing ground conditions and groundwater along with the interaction with other structures and construction activities. Section 3 describes the commonly used SCL construction methods. Along with the ICE report on SCL tunnels (ICE 1996), this offers a good starting point for understanding the limitations of the technique as well as potential hazards. Being an open face tunnelling method, SCL tunnels are more vulnerable to unstable ground (such as running sands) or instability caused by water inflows. Furthermore, spraying concrete onto this sort of ground will not cure the instability. Different measures are required such as ground treatment and forepoling or canopy tubes. As with any tunnelling method, adjacent structures – known ones like piles or unknown ones like abandoned wells – and adjacent activity – such as compensation grouting – can impose additional loads on the tunnel and affect its stability.

7.1.2 Materials

The hazards arising from the materials in the sprayed concrete can be handled in the normal way. The safety datasheets from materials suppliers and specifications (e.g. EFNARC 2002, BTS 2010) are a good starting point for information on these hazards. Most of the safety precautions are the same as the ones that one would normally apply to conventional concrete.

Generally speaking, the finished lining is inert and safe. However, one example where further steps may be required is the use of steel fibre reinforcement. The sharp fibres tend to protrude out of the surface of the lining. Where people may come into contact with this abrasive surface, a thin smoothing layer of plain concrete is often sprayed to prevent injury.

Besides normal construction risks, the major occupational health hazards associated with sprayed concrete come from the chemicals added and the dust created during spraying (see Section 3.4.1). The introduction of low alkali accelerators and wet mix spraying have done much to reduce these two risks to health respectively while robotic spraying has significantly reduced the manual handling during application.

7.1.3 Application and equipment

In SCL tunnels, the concrete is pneumatically sprayed at a high velocity onto the ground. This introduces hazards related to compressed air equipment and rebound as well as the normal ones associated with handling concrete and working underground. More information on safety during spraying can be found in published guidance such as EFNARC (1996), ITA (2008) and Lehto and Harbron (2011). Similarly, equipment suppliers also provide information on the safe operation of their equipment. Depending on the other support measures (such as bolting or lattice girders), a whole host of safety hazards may arise (see ITA 2004, ITA 2008) and should be addressed in the risk assessment.

As noted earlier, the concrete is sprayed as a liquid onto the ground, and, as it hydrates, the strength grows. In the first few hours after spraying, this introduces a risk that the loads from the ground and the self-weight of the concrete will cause localised collapses, occasionally with fatal consequences. Good workmanship is critical for safety in this area. One mitigation measure is to restrict temporarily worker access to areas of the sprayed concrete tunnel lining at risk. These areas are commonly referred to as "exclusion zones" (CLRL 2014). Typically, exclusion zones are required for a zone close to the tunnel face for tunnel excavation and the initial or primary lining installation, especially in the top heading and bench (above tunnel axis). Also, this may apply to zones anywhere along the tunnel where the upper arch of the final or additional sprayed concrete lining is under construction. Often "restriction zones" are also designated adjacent to "exclusion zones" to restrict non-essential personnel access by limiting access to those with designated roles and responsibilities for construction and quality inspection. The strength to be attained before re-entry should be defined in each case based on the specifics of the ground and the lining. Typically, the value is around 1.0 MPa (Rispin et al. 2017).

Safety has improved with the increasing automation and use of equipment in SCL tunnelling, such as robotic spraying and even remote methods for measuring the profile, convergence and strength.

7.1.4 Environmental impacts

Finally, regarding environmental protection, readers are advised to review their preferred usage of sprayed concrete in the light of the relevant local regulations and international best practice. SCL tunnelling does not normally add any major environmental hazards to those normally dealt with during tunnelling, such as storing and handling hazardous chemicals. There have been isolated instances of environmental pollution from SCL tunnels (e.g. Crehan et al. 2018). Sustainability is discussed in Section 2.1.9.

7.2 QUALITY CONTROL

The quality control regime for a tunnel is defined in the specification. Several countries have produced specifications for SCL tunnels – see Section 6.9 – which can be used like any other civil engineering specification. Full details of the test methods can be found in some of these specifications (e.g. EN 14487 2006, ACI 506.2 2013). EN 14487 is one of the most comprehensive and user-friendly specifications, so it has been used as the basis for many of the comments below. One peculiarity of SCL tunnels is that there two distinct phases of quality testing – pre-construction testing and testing during construction.

7.2.1 Pre-construction testing and staff competence

The purpose of pre-construction tests is simply to verify the suitability of the mix design and equipment before it is used in the actual tunnel. Typical tests are described in EN 14487 (2006) and Table 7.1. Durability tests such as water penetration or freeze–thaw should also be included when relevant.

Table 7.1 Pre-construction tests (EN 14487 2006)

Type of test	Category 2 Temporary support	Category 3 Permanent support
Consistency of the wet mix	Yes	Yes
Early age strength development	Yes	Yes
Compressive strength	Yes	Yes
Modulus of elasticity	Optional	
Bond to substrate	Optional	
Maximum chloride content	Optional	
And if using fibres		
Residual strength or energy absorption capacity	Yes	Yes
Ultimate flexural strength	No	Yes
First peak flexural strength	No	Yes

These tests should be completed at least one month before tunnelling starts so that it can be confirmed that the 28-day results meet the specified requirements. If a substantial change to the mix is planned during production, it will be necessary to repeat the pre-construction tests. On a note of caution, it should be remembered that mixes may perform differently in laboratory tests and full-scale spraying tests (e.g. Rudberg & Beck 2014) so both types of tests are needed.

Pre-construction tests also provide an opportunity to examine the competence of the construction crew. Despite the increasing automation, SCL tunnelling still relies heavily on the skill of key workers such as nozzlemen. Several international standards exist for assessing the competence of nozzlemen (e.g. ACI-C660 (2002) for dry mix and the EFNARC scheme for wet mix). While the focus traditionally has been on nozzlemen, pump operators too are important as they can influence the efficiency of spraying and the quality of the end-product. Goransson et al. (2014) describes a training programme for nozzlemen in detail, along with the use of a virtual reality training simulator.

This phase is a valuable opportunity to identify any weaknesses and remove them. The value of pre-construction testing cannot be over-emphasised. This effort can save much wasted time and money during construction.

7.2.2 Testing during construction

This section contains brief comments on the common tests. Full details of each test can be found in the references given. The frequency of testing should be commensurate with the scale of the works (e.g. EFNARC 1996 or see Tables 111 and 11/2 in ÖBV 1998, abbreviated in Appendix G, or EN 14487-1 Tables 9 to 12). The European standard, EN 14487, recommends

Table 7.2 Control of sprayed concrete properties (EN 14487 2006)

Type of test	Method	Category 2 Temporary support	Category 3 Permanent support
Control of fresh concrete			
1 w/c ratio	By calculation or test method	Daily	
2 Accelerator	From quantity added	Daily	
3 Fibre content	EN 14488-7	1/200 m³ or 1/1,000 m²	1/100 m³ or 1/500 m²
Control of hardened concrete			
4 Early age strength	EN 14488-2	1/2,500 m² or once per month	1/1,250 m² or twice per month
5 Compressive strength	EN 12504-1	1/500 m³ or 1/2,500 m²	1/250 m³ or 1/1,250 m²
6 Density	EN 12390-7	When testing compressive strength	
7 Bond strength	EN 14488-7	1/2,500 m²	1/1,250 m²
Control of fibre reinforced concrete			
8 Residual strength or energy absorption capacity	EN 14488-3 or EN 14488-5	1/400 m³ or 1/2000 m²	1/100 m³ or 1/500 m²
9 Ultimate flexural strength	EN 14488-3	When testing residual strength	
10 First peak flexural strength	EN 14488-3	When testing residual strength	
Fibre content can be measured when testing residual strength if not done already.			

minimum testing frequencies of inspection category 2 or 3 during normal production in tunnelling works (see Table 7.2).

A refinement of the regimes above could be to give the site team the option to reduce the frequency of testing if the results are consistently above the required values. Hauck (2014) describes the successful application of the approach for this defined in NCA (2011). Should the results deteriorate, the frequency could be increased again. In fact, EN 14487 recommends testing at four times the minimum frequencies stated above at the start of works or at critical sections.

Slump

This simple test gives an instant indication of whether the concrete is too stiff or too fluid. If the concrete is too stiff, it may not be possible to pump it so a high fluidity is desirable. Excessive fluidity may indicate that the water content is too high and the concrete may not adhere when sprayed. During spraying, cohesion and "stickiness" are desirable. For more fluid mixes, a flow table can be used instead. Typical slumps range from 180 to 220 mm while the spread on a flow table ranges from 500 to 550 mm. However, it is worth noting that these simple tests do not fully capture the complexity of the rheological situation and more sophisticated tests may be beneficial, especially during mix design (e.g. Thumann & Kusterle 2018, Yurdakul & Rieder 2018).

Compressive strength – early age

Typically the early age strength of sprayed concrete is tested at regular intervals (e.g. after one hour, six hours and 24 hours). Table 7.3 lists suitable test methods, and more information can be found in EN 14487 (2006).

Clements (2004) cautioned against the use of standard soil penetrometers for early age testing as they are prone to overestimating the strength. The cross-sectional area of the needle should be less than 10 mm². Thermal imaging cameras with the SMUTI system can be used to estimate the strength by means of the maturity of hydration (see Section 5.7.1) – see Jones et al. (2014) for more detail. The strength at an early age can also be measured indirectly by using ultrasound waves to measure the shear stiffness (Gibson & Bernard 2011, Lootens et al. 2014) which can be related to strength (e.g. Equation 5.1).

Compressive strength – mature

These tests are normally performed on cores from the lining itself or test panels. It may be necessary to repair core holes in the lining with a non-shrink mortar. Coring should not be used in secondary linings if there is a risk of puncturing a waterproofing membrane. Generally, cores can be taken once the sprayed concrete has reached a strength of 10 MPa (ÖBV 1998), although it may be possible to core at strengths as low as 5 MPa.

If necessary, the strengths measured from cored cylinders can be converted in cube strengths using established conversion factors. Typically, the cylinder strengths are around 80% of the cube strengths. In addition, some

Table 7.3 Test methods for sprayed concrete (ÖBV 1998)

Strength in MPa	Test method	Reference
0–1.2	Meyco or Proctor penetration needle	ASTM C403
1.0–8.0	Hilti gun penetration test	ÖBV 1998
3.0–16.0	Pull-out test on bolts	ÖBV 1998
16.0–56.0	Pull-out test on bolts	ÖBV 1998
> 10.0	Drilled cores	ÖBV 1998

have proposed a correction factor to allow for the disturbance caused by drilling. The Norwegian Concrete Association (NCA 1993) suggested the strength is reduced by a factor 0.8 compared to cast cubes.

Tensile strength

Tests on the direct tensile strength of sprayed concrete are rarely specified.

Flexural strength

A variety of test methods exist, using beams, round panels or square plates, to measure the flexural strength (see ITAtech 2016 for an overview). Bjontegaard and Myren (2011) contains a review of round and square panel tests. Unlike most of the other tests, these require special samples to be sprayed (rather than using samples from the lining), and the size of the samples is quite large. This increases the cost of testing. Concern has been voiced that there is an excessive variability of results from tests within one batch of beams (e.g. Collis & Mueller 2004). Notched beams (e.g. EN 14561) produce more consistent results but they also produce higher results because the location of the crack is pre-defined (ITAtech 2016). Round panel tests have been promoted as a more consistent alternative (Bernard 2004c).

Bond strength

Tests can be performed in accordance with EN 14488-4.

Rebound

The simplest method of measuring rebound is to lay a plastic sheet on the ground to collect the rebound as it falls. Knowing how much concrete was sprayed, the percentage of rebound can then be calculated. Typically rebound will be around 10 to 16% for wet mix and 21 to 37% for dry mix (Lukas et al. 1998).

Thickness

Traditionally thickness has been checked using markers such as thickness pins or by drilling holes into the lining. Now non-invasive methods are being used too, such as the DIBIT system (which takes 3D photographic images of surfaces), the Bever system (which uses 3D laser scanning) and the TunnelBeamer (which takes spot readings with a laser distometer) – see Figure 7.1. Currently, the spot measurement systems are the only ones which can be used in real-time while the concrete is being sprayed, which is the best way to control the thickness. However, by providing feedback to

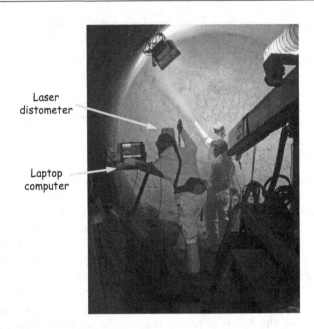

Laser distometer

Laptop computer

Figure 7.1 Laser-guided profile control.

the nozzlemen, the surface scanning systems can also be used to improve performance. Wetlesen and Krutrok (2014) report savings in the volume of concrete sprayed of up to 20%. Another major advantage of both systems is that they provide an as-built survey in an electronic format which could be integrated into a BIM model of the tunnel.

Where thin linings are in use, it may not be possible to obtain cores from the lining. In these cases, an area can be oversprayed to provide sufficient thickness for cores (typically more than 100 mm thick) or cores can be taken from test panels.

Durability

Various tests for durability may also be required depending on the purpose of the sprayed concrete – e.g. water penetration (EN 12390-8) or permeability tests and freeze–thaw resistance.

Some authors favour the use of tests for Boiled Water Absorption or Volume of Permeable Voids to assess durability. While these may be useful guides – for example, in the context of freeze–thaw resistance, they are not generally suitable as measures of durability (Jolin et al. 2011). Concrete strength and density are better indirect indicators of the *in-situ* durability of concrete.

When testing it is important to use representative samples. It is tempting to just perform tests where it is easy to do them (e.g. at the sidewall). Testing should be targeted to the areas where quality is most at risk (e.g. the crown where spraying overhead is more difficult or at joints where rebound may become trapped).

While it is undoubtedly important to perform quality control tests, designers should beware of specifying an unduly onerous regime. As the construction proceeds and the spraying operation emerges from the initial learning phase, there may be scope to reduce the testing frequency, if the specification is being consistently met. For example, since modern batching plants can provide very detailed records of the mix and *in-situ* tests can demonstrate that the required compressive strength has been achieved, it is arguable that fewer or no additional durability tests are needed during construction, unless the mix is changed.

7.3 INSTRUMENTATION AND MONITORING

During construction of an SCL tunnel it is normal to monitor the performance of the ground around the tunnel and the tunnel lining to ensure that it is behaving as expected. The review of monitoring data represents the umbilical cord that connects the growing construction with its design. In certain cases, visual observation augmented by a few simple monitoring sections may suffice (e.g. a simple tunnel in hard rock under an uninhabited area). In other cases, an extensive system of instrumentation may be installed in the ground and the tunnel (e.g. a metro station in soft ground under a city). Critical sections such as junctions should always be monitored carefully, and the designer should be involved in the review of the monitoring.

When specifying the monitoring regime, the frequency and sophistication of the monitoring should be appropriate to the scale and importance of the tunnel. However, the designer should also build in redundancy into the system since some instrumentation will inevitably be damaged or become unusable. All instruments should provide relevant information. Trigger values should be specified for the monitoring, based on the design. Figure 7.2 contains an example of a monitoring regime for a shallow urban tunnel – i.e. at the more rigorous end of the spectrum.

For certain instruments, it is important to obtain a set of baseline readings before the object in question is affected by the tunnel. This means that the effect of the tunnel can be clearly separated from seasonal effects or other influences.

7.3.1 Instrumentation

This section will focus on the monitoring of the sprayed concrete lining itself. An overview of instrumentation for the ground can be found in BTS Lining

Distance from face	Frequency	Distance from face	Frequency
Surface settlement points		Inclinometers & extensometers	
-30 to 0 m	Daily	-30 to -15 m	Twice Weekly
0 to +30 m	Daily	-15 to 0 m	Daily
+30 to +60 m	Twice Weekly	0 to +30 m	Daily
> +60 m	Weekly	+30 to +60 m	Weekly
		> +60 m	Weekly
Lining convergence points		Piezometers	
0 to +30 m	Daily	General	Weekly
+30 to +60 m	Twice Weekly	-15 to 0 m	Daily
> +60 m	Weekly	0 to +30 m	Daily
		> +30 m	Weekly

Figure 7.2 Monitoring regime for a shallow urban tunnel.

Design Guide (2004), van der Berg (1999), BTS (2011) and Chapman et al. (2017). This is a constantly evolving field with new instruments emerging; for example, fibre optics have been trialled for strain measurements in sprayed concrete linings (e.g. de Battista et al. 2015). A good description of monitoring of adjacent structures can be found in the compendium of papers on London's Jubilee Line Extension project (Burland et al. 2001). Table 7.4 and Table 7.5 list some of the instruments commonly used on tunnelling projects.

The normal hierarchy of monitoring is: in-tunnel convergence; surface settlement; subsurface instruments (e.g. inclinometers, extensometers, piezometers); in-tunnel stress–strain measurements. In other words, most weight is placed on the measurements of lining deformations, then surface movements, and so on.

In the sections below there are a few brief comments on the instruments and their usage.

Convergence monitoring

Tape extensometers are more accurate than optical surveying methods but taking measurements with the tape can be more disruptive since it blocks

Table 7.4 Instruments for monitoring ground behaviour

Settlement pins (surface)	Inclinometers	Earth pressure cells
Settlement pins (deep level)	Deflectometers	Piezometers
Electrolevels	Extensometers	

Table 7.5 Instruments for monitoring lining performance

Instrument	Typical range and accuracy
Convergence pins	(with tape extensometers) Up to 30 m and ± 0.003 to 0.5 mm (with 3D optical survey) ~100 m and ± 0.5 to 2.0 mm
Radial (Earth) pressure cells	0.35 to 5 MPa and ± 0.1% see Geokon website
Tangential (Shotcrete) pressure cells	2 to 20 MPa and ± 0.1 to 2.0%*
Vibrating wire strain gauges	Up to 3,000 με and ± 1 to 4 με

* This is the manufacturer's stated accuracy for the instrument alone; the accuracy when embedded in the lining is different – see Jones (2007).

traffic in the tunnel during the measuring. For this reason, optical surveying has largely replaced tape extensometers.

When interpreting the monitoring by 3D optical measurements, it is important to remember that the readings can be affected by changes in the atmosphere in the tunnel and the accidental disturbance of targets. Hence, sudden changes in readings may not actually indicate a change in deformation but merely that someone has knocked one of the pins.

Traditionally convergence monitoring has focused on the inward movement of the lining. However, recent research indicated that the longitudinal movements can provide useful information too, and they can be used to predict of changes in rock conditions (Steindorfer 1997).

Stress monitoring

Pressure cells are the traditional method of measuring stresses in SCL tunnels, although there are other methods (e.g. slot cutting and overcoring – see Jones (2007) for a full review). Many authors have questioned the reliability of pressure cells in sprayed concrete (e.g. Golser et al. 1989, Golser & Kienberger 1997, Mair 1998, Kuwajima 1999, Clayton et al. 2000), for the following reasons:

- The physical size of the cells (100 mm wide) may lead to shadowing. Incomplete encasement would lead to under-reading of stress.
- During the rapid hydration, the cell may expand and on cooling leave a gap between itself and the concrete, again leading to under-reading (Golser et al. 1989).
- The increase in readings of tangential cells due to thermal effects has been calculated as 0.10 MPa/°C for mercury-filled cells and 0.15 MPa/°C for oil-filled cells (Clayton et al. 2000).[1]
- The increase in readings of radial cells due to thermal effects has been calculated as about 0.06 MPa/°C for oil-filled cells (Jones 2007).

- Shrinkage can also induce stresses into the pressure cells (Clayton et al. 2000).
- The stiffness of the cell is different from the surrounding lining. If there is no difference the Cell Action Factor (CAF)[2] is 1.0. The CAF is often close but lower than one (Clayton et al. 2000, Jones 2005), leading to under-reading slightly.

The results may also depend on the measuring system. In one case, it was estimated that pressure cells with a hydraulic measuring system yielded readings that were about 80 kPa higher than vibrating wire cells (Bonapace 1997). Clayton et al. (2000) and Aldrian and Kattinger (1997) suggest that tangential cells record changes in stress accurately but should not be assumed to be recording the correct absolute values. The standard deviation in readings is often almost as large as the average readings themselves. On the Jubilee Line Project (JLE) the tangential stress was on average 2.0 MPa after three months (corresponding to about 25% of the full overburden pressure (FOB)) but ranged from 0.0 to 7.0 (Bonapace 1997).

Radial (Earth) pressure cells are believed to be more reliable because they are easier to install and the cell stiffness and the behaviour of the sprayed concrete have less influence on the readings (Clayton et al. 2000). However, like tangential cells, even when the results from a large number of cells are examined, there is usually considerable scatter in the results from radial cells (Bonapace 1997).

However, Jones (2007) has described how the errors listed above can be quantified and removed from pressure cell readings. The key steps are as follows:

- Install the pressure cells as soon as possible after excavation and minimise shadowing around the cells.
- Spray two tangential cells in a test panel at the same time as the main array, leaving the test panel to cure in the tunnel (i.e. in the same environment).
- Record readings at very frequent intervals during early hydration using a datalogger (e.g. ten minutes during first seven days and hourly thereafter).
- Crimp to ensure that the cells remain in contact with the concrete.

Crimping not only ensures that contact is maintained with the concrete but can also provide an indication of the quality of installation. Temperature sensitivity can also be used for this purpose. The post-processing of the readings requires the following procedure.

For tangential cells:

- Make adjustment for temperature sensitivity of the vibrating wire transducer (using manufacturer's calibration).

- Remove any zero offset.
- Remove any crimping offset.
- Check for lost pressures if the pressure cell has at any time lost contact with the sprayed concrete.
- Estimate cell restraint temperature sensitivity from test panel data, and estimate its variation with time during early age. Apply the correction for this.
- Estimate shrinkage pressure development with time from the test panel data and subtract from readings.

For radial pressure cells:

- Make adjustment for temperature sensitivity of the vibrating wire transducer (using manufacturer's calibration).
- Remove any zero offset.
- Remove any crimping offset.
- Check for lost pressures if the pressure cell has at any time lost contact with the sprayed concrete or ground. This is unlikely in the case of radial cells.

The correction for the overall temperature sensitivity for the embedded cell was found to be particularly important for tangential cells where even slight variations in ambient temperatures could induce changes in stress. This sensitivity explains some of the fluctuations in pressure cell readings. The correction for temperature sensitivity is complicated by the fact that the coefficient of thermal expansion changes with the age of the concrete (Jones et al. 2005). The apparent superior reliability of radial cells may simply be because they are not so prone to the effects of temperature and shrinkage.

This research seems to offer the important reassurance that – if the proper measures are taken – pressure cells can be used to obtain continuous and reliable measurement of the stresses in SCL tunnels. This enables the direct calculation of the factor of safety.

Finally, on the subject of stress measurements, a recent area of innovation is the use of software which estimates lining strains or stresses from lining deformations (e.g. Rokahr & Zachow 1997, Ullah et al. 2011). While interesting, it should be noted that there are some fundamental theoretical limitations in these methods (e.g. they are based on one dimensional constitutive models, or they assume that there is no bending in the lining). This may explain why some reports of these methods show peculiar results – such as zero or tensile axial forces (Hellmich et al. 2000, Ullah et al. 2011). Also, these calculations have rarely been calibrated against actual measurements of lining strains or stresses. Furthermore, Stark et al. (2002) note that considering the estimated stress intensity in the upper part of the lining can provide no warning of a failure in the invert, where deformation measurements are rarely taken.

Clayton et al. (2006) raised similar concerns. Finally, the variability in the readings of lining deformations (often ± 2 or 3 mm) and the relative infrequency of the measurements hamper these methods too (Jones et al. 2005). Consequently, it is arguable that, despite their increasing use, these systems have yet to be fully proven.

7.3.2 Trigger values

A system of trigger values is used to assess the performance of the tunnel. This section will describe one method that is commonly used.

Typically, there are three trigger values – a green, amber and red limit – see Figure 7.3. The green limit marks the boundary of normal behaviour. Crossing the amber limit indicates that there is a definite cause for concern while if the red trigger is crossed, tunnelling should stop. The red trigger should be set just below the ultimate capacity of the lining. The contractor's Action Plan should include pre-planned contingency measures that can be taken if a trigger value is exceeded – see also Powell et al. (1997) and BTS Lining Design Guide (BTS 2004).

The estimation of the trigger values can be summarised in the following procedure (see Figure 7.4). One should be careful when setting trigger values that are small and close to the size of the normal variation in the instruments' readings or when differences between triggers are of this size. Both carry the risk of causing many false alarms when the normal fluctuation in readings causes a trigger to be breached

There are alternative ways to set the trigger values but the rationale behind them is rarely documented. For example, one project set the amber trigger at 75% of the predicted settlement and the red trigger at 125%. The relationship between these values and the factor of safety was not stated. The method described in detail above has been used successfully on many major projects.

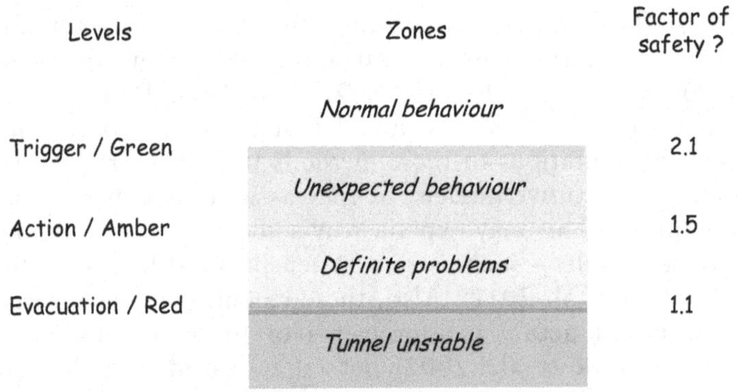

Levels	Zones	Factor of safety ?
	Normal behaviour	
Trigger / Green		2.1
	Unexpected behaviour	
Action / Amber		1.5
	Definite problems	
Evacuation / Red		1.1
	Tunnel unstable	

Figure 7.3 Trigger values.

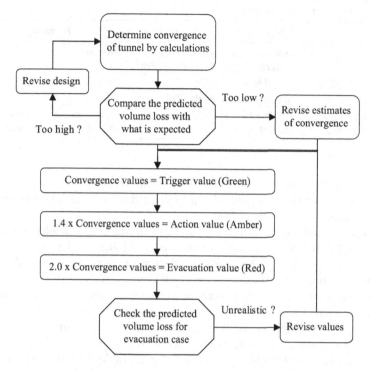

Figure 7.4 Derivation of trigger values.

Sometimes a hierarchy is specified for the interpretation of monitoring (e.g. see Section 7.3.1). Table 7.6 shows a typical hierarchy with the deformation of the lining at the top of it. The order of the ranking should reflect the key concerns of each project, as well as the reliability of the instruments. For example, if a project is installing pressure cells in the tunnel lining and it is believed that good results can be obtained from them, then arguably pressure cells should feature higher up in the ranking since they give direct information on the factor of safety of the lining.

Due to the accuracy of the instrumentation, a fluctuation in readings can be expected. Therefore, if an individual reading reaches a trigger value, it is

Table 7.6 Hierarchy for monitoring

Rank	Instrument/parameter
1	In-tunnel lining deformation
2	Surface settlement
3	Monitoring of adjacent structures (e.g. electrolevels, crackmeters)
4	Ground movement (e.g. inclinometers, extensometers)
5	Pressure cells (in-tunnel and in the ground)
6	Strain gauges

Table 7.7 Frequency of monitoring

Stage	Location of the tunnel face/timing	Frequency
Prior to tunnelling	At least three months before entering Zone of Influence (ZoI)	Weekly
During tunnelling	Tunnel face within ZoI	Daily
After tunnelling	Tunnel face outside the ZoI	Weekly
Until completion of monitoring	Not less than three months after leaving the ZoI	Three months or as instructed by the supervision team

not necessarily a cause for concern if the overall trend was stable; trends are more important than individual readings in determining whether trigger values have truly been exceeded.

When reviewing monitoring data, it is as important to consider the trends in the data as the absolute values themselves. For in-tunnel deformation readings, the tunnel is defined as stable when the rate of convergence has reduced to less than 2 mm per month. In high stress environments, the rate of convergence is sometimes used to judge the best time to install the final cast *in-situ* lining.

The frequency of monitoring is varied depending on how close the tunnel face is to the instruments and when the tunnelling will occur. The so-called "Zone of Influence" (ZoI) is used to delineate the areas which require the most frequent monitoring (see Figure 7.2 and Table 7.7). In shallow, soft ground tunnels, typically the zone of influence extends from three or four tunnel diameters ahead of the face to three or four tunnel diameters behind the closure of the completed ring.

The Zone of Influence also extends laterally from the centreline of the tunnel. The lateral extent may be defined according to the predicted settlement contours – e.g. set at a distance equal to the 1 mm contour.

7.4 DESIGNER'S REPRESENTATIVE ON SITE

The presence of a designer's representative (DR) on site provides continuity of the design through the construction phase. Typically, the DR will work alongside the supervision team, with the objective of promoting a single team approach. The exact relationship depends on the contractual environment of the project. The DR can assist in the following tasks:

- Review and approve the contractor's action plans (which define the monitoring regime, contingency measures and the criteria for implementing changes to support)
- Review and interpret monitoring data on a daily basis
- Review and approve changes in support and excavation sequences
- Advise on the impact of problems with the quality of construction

When required, the supervision team may issue instructions to the contractor on the basis of the DR's recommendations.

A single team approach, where key engineering decisions are made through assessment and evaluation of information on a routine basis during construction has clear benefits in terms of reducing risks related to safety and quality control. The single team approach does not mean that lines of responsibility are blurred as each organisation is required to appoint experienced staff to understand the engineering as well as contractual risks. This is arguably essential for the SCL approach where modifications and adaptations of the design within a specified framework are an inherent part of the design and construction process.

As an example, in one case the benefit–cost ratio of the presence of a DR was estimated at more than 3:1. The engineers helped to achieve cost savings of more than $300,000 (Thomas et al. 2003).

For small projects, a permanent presence on site may not be warranted but the designer should visit regularly and be involved in reviewing monitoring data. This is easy to achieve remotely with modern communications.

Alternatively, another way to communicate the design intent is to hold a workshop between the construction team and the designers before the works commence.

7.5 DAILY REVIEW MEETINGS

Monitoring alone serves no useful purpose. The construction of SCL tunnels is controlled by the process of reviewing the progress and monitoring data and taking actions based on that information, guided by criteria such as trigger values on deformations. Figure 7.5 illustrates the types of information that are fed into the review process. Daily review meetings (DRM) are standard practice. All parties actively involved with a tunnel should be represented at the DRM. For example, if compensation grouting is being performed nearby, the grouting team should be present too since their works affect the tunnel. The DRMs are augmented with Weekly and Monthly Review Meetings on more complex projects so that higher level staff can be kept informed on the overall performance of the tunnels. The data should be presented clearly at the meeting, ideally in graphs showing the absolute values and trends.

The outputs from the DRM are the instructions for the next day's tunnelling and any additional measures required to counter adverse trends in behaviour of the tunnels. An Excavation and Support Sheet is a simple tool to record these instructions and to communicate them to everyone concerned, from managers to the foreman at the tunnel face (see Figure 7.6).

If a trigger value is reached, first the site team should check that the reading is correct and consistent with the readings from other instruments. If the trigger has really been breached, then contingency measures can be

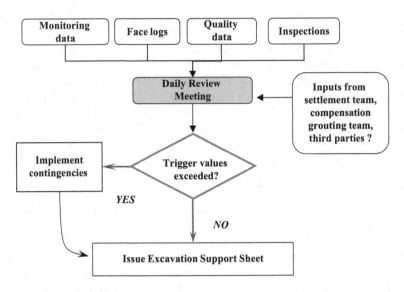

Figure 7.5 Information flow at the DRM.

Required Excavation & Support Sheet	
Site: ###	RESS no.: 1
Section: Vent Tunnel Advances 1 to 13	Chainage: #### to ####

Distribution	Senior SCL engineer, SCL shift engineer, pit boss, Resident Engineer, SCL inspector, Designer's Representative on site	
	Top Heading	Invert / Enlargement
Advance length	1.0 m	N/A
Reinforcement	Steel fibres (40 kg/m3)	
Initial layer	75 mm SFRS	N/A
Structural layer	150 mm SFRS	N/A
Finishing layer	Not applied yet	N/A
Other support	450 mm wide elephant's foot; 225 mm thick temporary invert	N/A

Remarks

For details refer to relevant construction drawings: 2154; 2156; 2157 & 2177

Before the start of any excavation, sufficient sprayed concrete must be available.

Construction tolerance is 75 mm; KWIK-A-STRIP starter bars to be installed as shown.

Install 3 monitoring targets at cross-sections of advances 4, 8 and the face of the headwall (5 no. targets in each).

Notes: SFRS – steel fibre reinforced sprayed concrete.

Constructor / SCL Engineer	Supervision Team / Designer
Prepared: ##.##.## 10:00	Approved: ##.##.## 10:00

Details of tunnel & chainages

Description of excavation & support sequence

Diagrams of excavation & support sequence & details

Additional notes

Signatures of designer, supervision team & constructor

Figure 7.6 Excavation and Support Sheet.

instigated, in accordance with a pre-defined Action Plan and as directed in the DRM. The contingency measures are designed to correct any anomalous behaviour. They range from increasing the frequency of monitoring and inspections (to gain a better understanding of what is happening), amending the excavation sequence, increasing the support measures or adding new support measures to, in the worst case, backfilling the tunnel. However, on a happier note, if these simple construction management techniques are followed, such drastic situations can be avoided easily.

NOTES

1. The temperature corrections quoted by the manufacturers of pressure cells normally refer to the transducer only and not the whole cell. The peak in recorded pressures often coincides with the peak in the temperature of the concrete (Clayton et al. 2000).
2. The Cell Action Factor (CAF) is the ratio of recorded pressure to actual pressure.

Appendix A

*The evolution of mechanical properties
of sprayed concrete with time*

Property & author	Equations	Constants	Range of applicability	Age range	References
Elastic modulus – Weber	$E_b = E_b^{28} \cdot a_e \cdot e^C$ $C = C_e/t^{0.60}$ t = age in days	$a_e = 1.084$ $C_e = -0.596$ $E^{28} = 30$ GPa; $a_e = 1.132$ $C_e = -0.916$ $A = 1.084$; $C_e = -0.916$ $E^{28} = 26$ GPa –Yin; $a_e = 1.084$; $C_e = -0.196$ – Huang (Typing error?) $a_e = 1.084$; $C_e = -0.596$ – Schropfer, Pottler	Z35F/45F cement Normal concrete Z25, 35L, 45L(Pottler) Used by Yin, Huang, Pottler and Schropfer		Alkhiami 1995; Yin 1996; Huang 1991; Schropfer 1995; Pottler 1990;
Compressive strength – Weber	$\beta_b = \beta_b^{28} \cdot a_\beta \cdot e^C$ $C = C_\beta/t^{0.55}$ t = age in days	$a_\beta = 1.27$ $C_\beta = -1.49$	Z35F/45F cement Normal concrete (Yin)		Alkhiami 1995; Yin 1996
Elastic modulus – Chang	$E = E^{28} \cdot a \cdot e^C$ $C = c/t^{0.70}$ t = age in days	$a = 1.062$ $c = -0.446$ $E = 3.86\, \sigma_c^{0.60}$	Based on literature review	$E - \sigma_c$ valid for 0–7 days; E valid for 4 hours to 28 days; $\alpha < 0.70$	Chang & Stille 1993; Chang 1994
Compressive strength - Chang	$\sigma_c = \sigma_c^{28} \cdot a \cdot e^C$ $C = c/t^{0.70}$ t = age in days	$a = 1.105$ $c = -0.743$ $\sigma_c = 0.105\, E^{1.667}$	Based on literature review	$E - \sigma_c$ valid for 0–7 days; valid for 4 hours to 28 days; $\alpha < 0.70$	Chang & Stille 1993; Chang 1994
Compressive strength – modified Byfors	$f_{cc} = 0.01 \times (A_1 t^{B1}/X) \cdot f_{cc}^{28}$ in % $X = 1 + (A_1/A_2) \cdot t^{(A1-B2)}$	Cylinder str in MPa, t = age in hours, For w/c = 0.4 – 0.5, $A1 = 0.300$ %;$A2 = 29.446$ % $B1 = 2.676$; $B2 = 0.188$ $f_{cc}^{28} = 29.022$ MPa	Back-calc from cores on panels sprayed in-situ		Kuwajima 1999

(Continued)

Property & author	Equations	Constants	Range of applicability	Age range	References
Elastic modulus – modified Byfors	$E_c = c_1 f_{cc}^{d1}/X$ in GPa $X = 1 + (c_1/c_2) \cdot f_{cc}^{(d1+d2)}$	$C_1 = 1306$ $C_2 = 7194$ $D_1 = 1.920; D_2 = 0.363$	Wide scatter for t < 10 hours		Kuwajima 1999
Elastic modulus – modified Byfors	$E_{cc} = 7.194 f_{cc}^{0.363}$		Strengths > 1 MPa		Kuwajima 1999
υ Poisson's ratio	$\nu_{cc} = \sigma f_{cc}^{b}$	$a = 0.128, b = 0.192,$ $r = 0.847, n = 52$ and $\dfrac{f_{cc}}{f_{cc}^{28d}} = at^b$	Strengths > 1 MPa		Byfors 1980
Elastic modulus – Golser	$E_t = E_{28} \cdot ([a + bt]/t)^{-0.5}$	$a = 4.2; b = 0.85;$ t in days			Yin 1996
Compressive strength – Aldrian	$\beta_t = \beta_{28} \cdot 0.03t$ $\beta_t = \beta_{28} \cdot [(t-5)/(45+0.975t)]^{0.5}$	$t = < 8$ hours $t > 8$ hours	For prisms; $\beta_{cube} = 1.07 \beta_{prism}$ for the sizes used by Aldrian		Aldrian 1991
Elastic modulus – Aydan	$E_t = A(1 - e^{Bt})$	$A = 5000, B = -0.42$	Wet mix, loaded at right angles to spraying	Age = 3 hours to 28 days	Yin 1996; Aydan et al. 1992a
υ - Poisson's ratio - Sezaki	$\nu_t = a + b.e^{ct}$	$a = 0.18, b = 0.32,$ $c = -5.6$	Wet mix, loaded at right angles to spraying	Age = 3 hours to 28 days	Yin 1996; Aydan et al. 1992

(Continued)

Property & author	Equations	Constants	Range of applicability	Age range	References
Compressive strength – Meschke	$f_{cu(t)} = f_{cu(1)} [(t+0.12)/24]^{0.72453}$ $f_{cu(t)} = ac \cdot e^{(-bc/t)}$ $ac = 1.027.fcu28$ $bc = 17.80$	for t < 24 hours for t > 24 hours, where $K = 0.489 = f_{cu1}/f'_{cu}{}^{28}$; ac & bc are functions of K	Austrian shotcrete guidelines for t < 24 hours		Meschke 1996; Kropik 1994
Elastic modulus – Meschke	$E_t = \beta_{Et} \cdot E_{28}$ $\beta_{Et} = 0.0468t - 0.00211\, t^2$ $\beta_{Et} = (0.9506 + 32.89/[t-6])^{-0.5}$	t < 8 hours 8 < t < 672 hours	Modified CEB-FIP	Age = 0 to 28 days	Meschke 1996, Kropik 1994
Compressive strength – Pottler	$\beta_t = \beta_1 . t^{0.72453}$ or $\beta_t = \beta_1 . (1 + 4t)/5$	$\beta_1 = 1$ day str	From cube tests.	Age > 1 day	Pottler 1993; Pottler 1990
Compressive strength – Eierle & Schikora	$f_{cc}(\alpha)/f_{cc,\alpha=1}$ $= [(\alpha - \alpha_0)/(1 - \alpha_0)]^{3/2}$	$f_{cc,\alpha=1}$ = compressive strength at complete hydration; α = degree of hydration $\alpha_0 = 1.8$ est.	For plain concrete; for tensile strength, $f_{ct}(\alpha)/f_{ct,\alpha=1}$ $= (\alpha - \alpha_0)/(1 - \alpha_0)$		Eierle & Schikora 1999
Elastic modulus – Eierle & Schikora	$E(\alpha)/E_{\alpha=1}$ $= [(\alpha - \alpha_0)/(1 - \alpha_0)]^{2/3}$	$E_{\alpha=1}$ = modulus at complete hydration; α = degree of hydration			
Compressive strength – Kusterle	Prisms $Y = 1.8171X^{0.8285}$; Cylinders $Y = 0.0521X^2 + 0.3132X + 0.9213$; Beams $Y = 0.9208X^{0.9729}$	Y = prism, cylinder or beam X = cube str in MPa Cube = 20x20x20 cm; Prism = 4x4x16 cm; Cylinder = 10 cm dia.x10 cm; Beam = 10x10x50 cm	Str range = 1 to 12 MPa		Testor & Kusterle 1998

NB: (see Appendix C Part 3 for tensile strength).

Appendix B

Nonlinear elastic constitutive model for sprayed
concrete (after Kostovos & Newman 1978)

$$G_{tan} = \frac{G_0}{\left(1+\left(C.d.\left(\frac{\tau_0}{f_{cyl}}\right)^{d-1}\right)\right)} \tag{B.1}$$

$$K_{tan} = \frac{K_0}{\left(1+\left(A.b.\left(\frac{\sigma_0}{f_{cyl}}\right)^{b-1}\right)-\left(k.l.m.e\left(\frac{\tau_o}{f_{cyl}}\right)^{n}\right)\right)} \tag{B.2}$$

where

$$e = \frac{\left(\frac{\sigma_0}{f_{cyl}}\right)^{m-1}}{\left(1+\left(l.\left(\frac{\sigma_0}{f_{cyl}}\right)^{m}\right)\right)^2} \quad \text{if } s_0 > 0.0, \text{ otherwise } e = 1.0 \tag{B.3}$$

and

f_{cyl} = the uniaxial compressive cylinder strength in MPa
G_{tan} = the tangent shear stiffness in MPa
G_o = the initial shear stiffness in MPa
K_{tan} = the tangent bulk stiffness in MPa
K_o = the initial bulk stiffness in MPa
σ_0 = octohedral mean stress in MPa = $(\sigma_1 + \sigma_2 + \sigma_3)/3$
τ_0 = octohedral shear stress in MPa = $\{\sqrt{((\sigma_1-\sigma_2)^2+(\sigma_2-\sigma_3)^2+(\sigma_3-\sigma_1)^2)}\}/3$

If $f_{cyl} \leq 31.7$ MPa, A = 0.516

Otherwise, $A = \dfrac{0.516}{\left[1+0.0027\cdot\left(f_{cyl}-31.7\right)^{2.397}\right]}$

$b = 2.0 + \left(1.81\,e^{-08}\cdot\left(f_{cyl}^{4.461}\right)\right)$

If $f_{cyl} \leq 31.7$ MPa, $C = 3.573$

Otherwise, $C = \dfrac{3.573}{\left[1 + 0.0134 \cdot \left(f_{cyl} - 31.7\right)^{1.414}\right]}$

If $f_{cyl} \leq 31.7$ MPa, $d = 2.12 + \left(0.0183 * f_{cyl}\right)$

Otherwise, $d = 2.70$

If $f_{cyl} \geq 15.0$ MPa, $k = \dfrac{4.000}{\left[1 + 1.087 \cdot \left(f_{cyl} - 15.0\right)^{0.23}\right]}$

Otherwise, $k = 4.000$

$$l = 0.22 + \left(0.01086 * f_{cyl}\right) - \left(0.000122 * f_{cyl}^2\right)$$

If $f_{cyl} \leq 31.7$ MPa, $m = -2.415$
Otherwise, $m = -3.531 + (0.0352 * f_{cyl})$
If $f_{cyl} \leq 31.7$ MPa, $n = 1.0$
Otherwise, $n = 0.3124 + (0.0217 * f_{cyl})$

Other notes: Kotsovos & Newman's model is formulated in terms of octahedral stresses, normalised by division by the uniaxial cylinder strength of the concrete. For height to diameter ratios of 2.0 to 2.5, the uniaxial cylinder strength is approximately equal to the uniaxial cube strength (Neville 1995); the relationships provided by Kotsovos & Newman for predicting the initial bulk and shear moduli have not been used, since they are valid for mature concrete. The model has also been extended to apply to concrete at strengths of less than 15 MPa.

Appendix C

Plasticity models for sprayed concrete

PART I PLASTICITY (& NONLINEAR ELASTICITY) MODELS FOR COMPRESSIVE STRESS REGION

Author	2D/3D	Elastic region	Yield criterion	Post-yield	Notes	Visco-plasticity	Shrinkage
Meschke, Kropik	3D	Time-dependent elastic modulus	Drucker–Prager	Strain hardening, with limit strain dependent on age	Ultimate strength dependent on time; constant yield / ultimate strength ratio; biaxial; isotropic	Yes – Duvant– Lions formulation	?
Hellmich	2D	Hyperelastic	Drucker–Prager	Strain hardening	Potential damage levels assessed; isotropic	No	No
Lackner	2D	?	Drucker–Prager	Strain hardening	Isotropic	No	No
Hafez	3D	Linear elastic ?	3 parameter Chen–Chen model	Strain hardening	Post peak strain softening could be included; multiple SCL layers; isotropic	No	No
Haugeneder	3D	Time-dependent elastic modulus	Buyukozturk model	?	Reinforcement smeared across elements but assumed isotropic?	No	No
Moussa, Aydan	2D	Nonlinear elastic model (implicitly) with time-dependent elastic modulus	Strain history model	Strain hardening	Damage effects included; post peak strain softening could be included; isotropic; reinforcement incorporated using effective areas	No	No

(Continued)

Author	2D/3D	Elastic region	Yield criterion	Post-yield	Notes	Visco-plasticity	Shrinkage
Hellmich et al.	2D	Thermo-chemo-mechanical coupling to account for ageing	Drucker–Prager	Strain hardening?	Associated flow rule	No	Yes – but autogeneous shrinkage only
Watson et al.	2D ?	Linear elastic E=7.5 GPa for t < 10 days, then E = 15 GPa	?	Perfectly plastic	Yield stress = 5 MPa for t < 10 days, then yield stress = fcu/1.5; ultimate strain calculated using Rokahr & Lux's Kelvin rheological model	No	No

PART 2 PLASTICITY (& NONLINEAR ELASTIC) MODELS FOR TENSILE STRESS REGION

Author	2D/3D	Yield criterion	Post-yield	Type of crack model	Notes
Meschke, Kropik	3D	Rankine	Linear softening	Rotating crack	Exponential softening also possible
Lackner	2D	Rankine	Linear softening, exponential or both	Fixed and rotating crack models investigated	Anisotropic; damage model
Hafez	3D	Chen–Chen	Strain hardening, brittle failure	–	Extensive of compression model
Haugeneder	3D	Composite	Curve fitted to match tension stiffening effect of reinforcement	?	Reinforcement smeared across elements; isotropic?
Moussa	2D	Extended compression model	Linear softening	Smeared crack	Nonlinear elastic model
Hellmich et al	2D	Tension cut-off	–	–	Drucker–Prager surface up to cut-off

PART 3 PARAMETERS FOR PLASTICITY MODELS

Author	Age	Compressive yield/ ultimate strength	Comp. yield strain %	Comp. ultimate strain %	Tensile ultimate strength	Tensile yield strain %	Tensile ultimate strain %
Meschke / Kropik (1. see below)	t < 8 hours 8 < t < 16 t > 16 hours	—	—	6–0.25 0.25–0.17 0.17–0.20	$0.0067.fcu^{1.09}$ (fcu in kN/cm^2)	—	—
Meschke (ACI)	28 days	0.1–0.25	—	—	$0.32.fcu^{0.5}$	—	Softening gradient, $Dt = Et/100$
Lackner	?	0.4	—	—	$0.3fcu^{0.67}$	—	—
Hafez	28 days	0.3	—	—	$0.1.fcu$	Linear up to 0.6 ftu	—
Moussa / Swoboda	28 days?	0.3	Strain @ peak stress, ε_l 0.3 – 0.5 – 1.0	0.5–1.0	$0.21–0.3.fcu^{0.67}$	$\varepsilon_{cr} = \varepsilon_l[1 -v(1 -(f_t/f_c))]$	0.03
Aydan et al.	12 hours	—	—	3	—	—	—
BS 8110 Part I 1997	?	—	$0.24 (fcu)^{0.5}$ (@ peak)	0.35	—	—	—

1.Formulae for strain at compressive failure, t < 8 hours, $\varepsilon = 0.06 – [0.0575.t /8]$; $8 < t < 16$, $\varepsilon = 0.0025 – [0.0008.(t – 8)/8]$; $t > 16$ hours, $\varepsilon = 0.0017 + [0.003 .(t – 16)/634]$

Appendix D

Creep models for sprayed concrete

Author	Equations	Constants	Range of applicability & notes	Type	References
Rheological models					
Rokahr & Lux	$\Delta\varepsilon_v / \Delta t = \left[(1/2\eta_{k\;ov\;ta})\right]$ $\cdot\left(1-\left\{\varepsilon_v^{v,tr}\cdot 3G_{k(ov)/ov}\right\}\right)\cdot M\cdot\sigma$ ta = age of shotcrete $\varepsilon_v^{v,tr}$ = accumulated transient inelastic deformation (eff. strain) – $\varepsilon_v^v = \left(2^{0.5}/3\right)\cdot\left[\left[\varepsilon_{1v}-\varepsilon_{2v}\right]^2\right.$ $\left.+\left[\varepsilon_{2v}-\varepsilon_{3v}\right]^2+\left[\varepsilon_{3v}-\varepsilon_{1v}\right]^2\right)^{0.5}$ σ_v = eff stress = $(3J_2)^{0.5}$ M = condensation matrix ε_v = integral over time of viscous strain rate $\Delta\varepsilon_v/\Delta t$ = viscous strain rate	$3G_k = G_k^*\cdot e^{k1\sigma v}$ $3\eta_{lk} = \eta_k^*\cdot t_a^n\cdot e^{k2\sigma v}$ At 8 hours age: $G_k^* = 8450$ MPa, $\eta_k^* = 6900$ day.MPa, $k1 = -0.459$ MPa^{-1}, $k2 = -0.642$ MPa^{-1}, $n = 0.74$	8 hours to 10 days (?); stress < 10 MPa; At 24 hours age: $G_k^* = 52500$ MPa, $\eta_k^* = 17000$ day.MPa, $k1 = -0.0932$ MPa^{-1}, $k2 = -0.100$ MPa^{-1}, $n = 0.73$ (Berwanger's parameters)	Visco-elastic Modified Kelvin	Rokahr & Lux 1987; Watson et al. 1999
Schropfer	See Rokahr &Lux	At 30 hours age: $G_k^* = 1312$ MPa (?), $\eta_{lk}^* = 17315$ day.MPa, $k1 = -0.0309$ MPa^{-1}, $k2 = -0.2786$ MPa^{-1}, $n = 0.70$	Time = 0 to 30 hours, Assumes const.Volume during creep, Assumes hydrostatic stress has no influence on creep; B35 shotcrete	Visco-elastic Modified Kelvin	Schropfer 1995

(Continued)

Author	Equations	Constants	Range of applicability & notes	Type	References
Mertz	(see Schropfer)	Not given	—	Visco-elastic Modified Kelvin	Schropfer 1995
Kuwajima	$\varepsilon_c(t')/_{\sigma 0} = 1/E_\tau \cdot (1 - e^{-\lambda t'})$ $\lambda = E_\tau/\eta$	$E = 15$ GPa $1/E_\tau = 0.03$ GPa^{-1} $\lambda = 0.003$ min^{-1} NB: average values	Age = 10 to 100 hours; time = 10 to 100 hours; Notes that adding time dependency might overcome some inaccuracy due to use of average values.	Visco-elastic Generalised Kelvin unit	Kuwajima 1999
Swoboda	$\varepsilon_c(t')/_\sigma = 1/E_{b\,ta}(1 - e^{-\lambda t'})$ $\lambda = E_{b\,ta}/\eta$ $\varepsilon = _\sigma/E_b + \varepsilon_c$	t < 2 days: $E_b = 15$ GPa, $E_{b\,ta} = 6$ GPa, $\eta = 6 \times 10^8$ day kPa t > 2 days: $E_b = 25$ GPa, $E_{b\,ta} = 10$ GPa, $\eta = 10 \times 10^8$ day kPa	—	Visco-elastic Generalised Kelvin unit	Swoboda & Wagner
Petersen	$\Delta\varepsilon_{ij}^v / \Delta t = \left[(1/3\eta_k \cdot \sigma_{vt})\right.$ $\left. \cdot (e^{A \cdot t}) \cdot (3/2) \cdot M_2 \cdot \sigma_{ij}\right]$ $A = -3G_{k(\sigma v)}/3\eta_{k(\sigma v,t)}$	$3G_k = G_k^* \cdot_\sigma^{ek1\,v}$ $3\eta_k = \eta_k^* \cdot t^n \cdot e_\sigma^{k2\,v}$ $\Delta\varepsilon_v/\Delta t$ = viscous strain rate	30 hours to 10 days and $\sigma_v < 12$ MPa, $G_k^* = 1312$ MPa, $\eta_k^* = 17312$ day.MPa, k1 = -0.0309, k2 = -0.2786, n = 0.70	Based on modified Burger's model	Yin 1996; Pottler 1990

(Continued)

Author	Equations	Constants	Range of applicability & notes	Type	References
Pottler	$\Delta \varepsilon_{ij}^{v} / \Delta t = \left[\alpha_t \cdot \sigma_v^2 \right.$ $\left. + \beta_t \cdot \sigma_v^3 \right] \cdot (3/2) \cdot M_2 \cdot \sigma_{ij}$ $\alpha_t = (0.02302 - 0.01803t$ $+0.00501t^2).10^{-3}$ $\beta_t = (0.03729 - 0.06656t +$ $0.02396t^2).10^{-3}$	(see Petersen) $\upsilon = 0.167;$		Polynomial form of Petersen	Yin 1996; Pottler 1990
Yin	$\Delta \varepsilon_{ij}^{v} / \Delta t = \left[\left(a_t \cdot \sigma_v^{(b-1)} \right) \cdot (3/2) \cdot M_2 \cdot \sigma_{ij} \right]$ $a_t = A1.e^{(A2/(A3+t))}$ $b_t = B1.e^{(B2/(B3+t))}$ See also el.Vis. pl. rock model with DP and strain softening	$\upsilon = 0.2,$ $A1 = 0.20167 \times 10^{-5},$ $A2 = 3.9444,$ $A3 = 0.758,$ $B1 = 1.4791,$ $B2 = 0.15983,$ $B3 = 0.602$	Stress up to 20 MPa +, Duration up to 20 days +	Based on Petersen but with age dependent creep	Yin 1996
Huang	$\Delta \varepsilon_{ij} = \Delta \varepsilon_{ij}^{E} + \Delta \varepsilon_{ij}^{I} + \Delta \varepsilon_{ij}^{\mu}$ $\Delta \varepsilon_{ij}^{E} = E_{ijkl}(t + \Delta t) \Delta \sigma_{kl} + \Delta E_{ijkl}^{-1} \cdot \sigma_{kl};$ $\Delta \varepsilon_{ij}^{I} / \Delta t = S_{ij} / \eta_{ln(t,\sigma v)}; \Delta \varepsilon_{ij}^{II} = \Sigma_{\mu} \Delta \varepsilon_{ij}^{\mu}$ $\Delta \varepsilon_{ij}^{\mu} = \left(S_{ijkl} \cdot \sigma_{kl} \cdot (t + \Delta t) / E_{\mu} \right)$ $- \varepsilon_{ij}^{\mu} \left(1 - e^{A} \Delta^t \right),$ $A = -E_{\mu} / \eta_{\mu},$ $\varepsilon_{ij}^{\mu} =$ sum of strains to t from $t = \tau_o,$ $\upsilon = 0.2, \tau_o = 0.3$ days	Spring, $E_{b28} = 34$ GPa see also below, Dashpot calculated using Berwanger's values at 8 hours; Kelvin units 1 & 2 – $E_1 = 1.0 \times 10^4$ MPa, $\eta_1 = 1.0 \times 10^5$ MPa / day, $E_2 = 1.4 \times 10^4$ MPa, $\eta_2 = 1.4 \times 10^6$ MPa / day	As for Rokahr & Lux	Visco-elastic (Spring in series with dashpot and 2 Kelvin units) Time-dep E	Huang 1991

(Continued)

Author	Equations	Constants	Range of applicability & notes	Type	References
Zheng	(see Yin)	(see Yin)	Limited to low stress	Spring in series with dashpot and 2 Kelvin units	Yin 1996
Brite Euram	Strain rate, $\dot{\varepsilon}_{ij}' = \varepsilon_{ij}'^{e} + \varepsilon_{ij}'^{vp}$ $\dot{\varepsilon}_{ij}' = -(\upsilon/E)\dot{\sigma}_{kk}'\delta_{ij} + (1+\upsilon)\dot{\sigma}_{ij}'/E$ $\varepsilon_{ij}'^{vp} = (F/2\eta)\sigma_{ij}''$ F = yield function $= 1 - K/(0.5\sigma_{ij}'' \cdot \sigma_{ij}'')^{0.5}$ σ_{ij}'' = deviatoric stress $= \sigma_{ij} - (1/3)\sigma_{kk}\delta_{kk}$ $K = A\sigma_h + B$ σ_h = hydrostatic stress	At age = 4 hours, E = 255 MPa η = 2145 MPa/s K = 2.06 MPa (A = −14, B = 0.7 for K = 9.1 MPa)	E, K, η available for age t = 1 to 72 hours; Stress = 0 to 30 MPa? At age = 72 hours, E = 4865.24 MPa η = 2966.7 MPa/s K = 18.34 MPa	Elastic visco-plastic (in shear) Bingham model	Brite Euram C1 1997

Power creep law models

Author	Equations	Constants	Range of applicability & notes	Type	References
Alkhiami	$\varepsilon = \varepsilon^{el} + \varepsilon^{kr}$ $\varepsilon^{kr} = k \cdot \sigma_{eff}^{n} \cdot t^{m}$ σ_{eff} is effective stress	m = 0.252 n = 2.574 k = 1.88 × 10⁻⁵ (?)	Large exc.; Range of m for stress = 5 to 9 MPa and n for 0.1 to 168 hours Creep law in Abaqus	Straub 1931	Alkhiami 1995

(Continued)

Author	Equations	Constants	Range of applicability & notes	Type	References
Rathmair	$\Delta\varepsilon / \Delta t = A \cdot \sigma_{eff}^n \cdot t^m$ (is this formula correct? – see Alkhiami) – see also Schubert's Rate of Flow Method	m = -0.252, n = 2.574, A = 1.88 × 10⁻⁵	Doesn't handle elastic strain well; transverse strains don't agree well; Creep law in Abaqus		Rathmair 1997; Golser 1999;
Probst	$\varepsilon = A \cdot \sigma_{eff}^n \cdot t^m$	m = 0.36, n = 0.75, A = 1.12 × 10⁻⁴	Stress range = 2.5 – 10 MPa; age range = 1 to 100 hours		Probst 1999
Rate of flow Aldrian-modified Schubert model	$\varepsilon_2 - \varepsilon_1 = (\sigma_1-\sigma_2)/(E_{28}\cdot V^*\cdot f) + \sigma_2\Delta C.$ $V^* = (1/22.5)\cdot[25\cdot(1-\alpha)\cdot(t/\{25+1.2t\})^{0.5} + 3\alpha(t/\{100+0.9t\})^{0.5}]$ Viscous deform. $\Delta C = A(t - t_1)^X$, $X = 0.25.\alpha^{0.2}$ (or simply X = 0.25); Delayed elastic increment, $\Delta\varepsilon_d = (\sigma_2 C_{d00} - \varepsilon_d)\{1 - e^C\}$, $C = -\Delta C/\theta$; $\varepsilon_{sh} = \varepsilon_{sh,00}\cdot t/(B + t)$; Temp. deform, $\varepsilon_t = 30.$ $[1 - \cos(250t^{0.25})]$	ε in 10⁻⁶; t ln days, except for ε_t; V* = Relative deformation modulus; α = utilisation factor; unload constant; f = 1.0; A = constant, where A = 0.25 × 10⁻³, t = 0; A = 0.04 × 10⁻³, t = 500 hours; C_{d00} = limiting value of rev. creep = 0.0001, A/θ = 10; $\varepsilon_{sh,00}$ = limiting value of shrinkage = 1200με, B = 35 (ACI 1978) to 70 hours (Probst);	Age = 1 to 14+ days; Dry mix; for 0 < α <1.0 though behaviour may be different for α >0.8; Aldrian supersedes Schubert model; the temp. deform only applies for 0 < t < 4 days; Unloading constant may vary between 1.1 and 1.5 (Probst 1999)		Golser et al. 1989; Golser 1999; Schubert 1988; Aldrian 1991; Probst 1999
Schubert original model	Plastic deform. $C_t = At^{0.33} \cdot e^{k\alpha}$ for α > 0.5; $C_t = At^{0.33}$ for α <0.5		Power law		Schubert 1988;

Appendix E

Key to figures from Thomas (2003)

PART 1 SPRAYED CONCRETE CONSTITUTIVE MODELS IN THE NUMERICAL STUDY

Abbreviation	Description – see also Chapter 5*
E-	Linear elastic, constant stiffness = 28 day value
Et	Linear elastic, age-dependent stiffness – see 5.1
HME	Hypothetical Modulus of Elasticity – see 5.2
MCSS	Strain-hardening plasticity model (Mohr–Coulomb) – see 5.3.2
NLE	Nonlinear elastic model - Kotsovos & Newman (1978) – see 5.3.1
VE 1	Visco-elastic "Kelvin" creep model – stress independent – see 5.6.2
VE 2	Visco-elastic "Kelvin" creep model – stress dependent – see 5.6.2
VE 3	Visco-elastic "Kelvin" creep model after Yin (1996) – see 5.6.2

* The cross-references refer to sections in the original thesis.

PART 2 NUMERICAL MODEL RUNS FOR TUNNEL EXCAVATION

Run	Key feature – see also Chapters 4 & 6*
H series	**Exact geometry of HEX Platform tunnels**
H_Et_4	BASE CASE – Linear elastic, age-dependent stiffess for lining; ground model 4 = strain-hardening plasticity
H_E-_4	Linear elastic, age-independent stiffness
H_HME_4	Hypothetical Modulus of Elasticity
H_MCSS_4	Strain-hardening plasticity model (Mohr–Coulomb)
H_NLE_4	Nonlinear elastic model after Kotsovos & Newman (1978)
H_VE1_4	Visco-elastic "Kelvin" creep model – stress independent
H_VE2_4	Visco-elastic "Kelvin" creep model – stress dependent

Run	Key feature – see also Chapters 4 & 6*
H_MC_4_JR	MCSS model but strength reduced by 50% on radial joints
H_MC_4_JL	MCSS model but strength reduced by 50% on longitudinal joints
H_MC_4_J	MCSS model but strength reduced by 50% on radial & long. Joints
H_Et_4_A_0.5	Base case with advance length = 0.5m
H_Et_4_A_2.0	Base case with advance length = 2.0m
H_MC_4_A_2.0	H_MC_4 with advance length = 2.0m
H_Et_4_X_6.0	Base case with average ring closure distance of 6.0m
H_Et_4_X_8.0	Base case with average ring closure distance of 8.0m
H_Et_0	Ground model 0 = linear elastic
H_Et_1	Ground model 1 = linear elastic perfectly plastic – spft
H_Et_2	Ground model 2 = anisotropic linear elastic
H_Et_3	Ground model 3 = nonlinear anisotropic elastic
H_HME_3	As per H_Et_3 with HME lining model
H_Et_5	Ground model 5 = linear elastic perfectly plastic – stiff
H_Et_4_K_1.5	Constant K_0 = 1.50
H_Et_5_K_1.5	Constant K_0 = 1.50
N series	**Circular tunnel with face area equivalent to HEX Platform tunnel**
N_Et_4	Base case – see H_Et_4
N_E-_4	Linear elastic, age-independent stiffness
N_HME_4	Hypothetical Modulus of Elasticity
N_MCSS_4	Strain-hardening plasticity model (Mohr–Coulomb)
N_VE2_4	Visco-elastic "Kelvin" creep model – stress dependent
N* series	**Circular tunnel with face area equivalent to HEX Concourse tunnel** **NB: K0, advance rate, advance length & ring closure distance differ from H_Et_4**
N*_Et_4	Base case – see H_Et_4
N*_E-_4	Linear elastic, age-independent stiffness
N*_HME_4	Hypothetical Modulus of Elasticity
N*_MCSS_4	Strain-hardening plasticity model (Mohr–Coulomb)
N*_VE1_4	Visco-elastic "Kelvin" creep model – stress independent
N*_VE3_4	Visco-elastic "Kelvin" creep model – after Yin (1996)
N*_Et_4_S_1.2	Base case with advance rate of 1.2 m /day
N*_Et_4_S_0.5	Base case with advance rate of 0.5 m /day
N*_Et_4_X_4.5	Base case with average ring closure distance of 4.5m
N*_Et_0	Ground model 0 = linear elastic
N*_Et_1	Ground model 1 = linear elastic perfectly plastic – spft
N*_Et_2	Ground model 2 = anisotropic linear elastic
N*_Et_3	Ground model 3 = nonlinear anisotropic elastic
N*_Et_5	Ground model 5 = linear elastic perfectly plastic – stiff
N*_Et_4_K	K_0 profile varies with depth

* The cross-references refer to sections in the original thesis.

PART 3 GEOTECHNICAL MODELS *

Model 0   Linear elastic isotropic model

Cu = 0.67*(50+(8*depth below surface))   in kPa
(the 0.67 factor is to convert values from laboratory tests to a mass
  property)
E = 600*Cu   in kPa
Equivalent to the stiffness at a deviatoric strain of 0.1 %
$v = 0.49$
K = 100*G   for undrained case

Sources: Mott MacDonald 1990

Model I   Linear elastic perfectly plastic isotropic model

Cu, E, v, K – as per Model 0, for undrained case
Mohr–Coulomb (Tresca) yield criterion
Tension limit, dilation and friction for plastic model all set to zero.

Model 2   Linear elastic transversely anisotropic model

Cu – as per Model 0, for undrained case

E_{vu} = 600*Cu   (vertical stiffness)   in kPa
E_{hu} = 1.6*E_{vu} (stiffness in horizontal plane)   in kPa
G_{vh} = 0.433*E_{vu} in kPa
v_{hh} = 0.20
v_{vh} = 0.48   (as per Lee & Rowe 1989 for undrained case)

Sources: Van der Berg 1999, Lee & Rowe 1989

Model 3   Nonlinear elastic tranversely anisotropic model

Cu, E_{hu}/E_{vu}, G_{vh}/E_{vu}, v_{hh}, v_{vh} – as per Model 2 for undrained case
The overconsolidation ratio, OCR, varies with depth.

If ε_{dev} < $1.0e^{-0.5}$,
   G_{vh} = $0.828.(OCR)^{0.2}.p'.(0.00001^{-0.501})$ in kPa
If 0.01 > ε_{dev} > $1.0e^{-0.5}$,
   G_{vh} = $0.828.(OCR)^{0.2}.p'.(\varepsilon_{dev}^{-0.501})$ in kPa
If 0.01 < ε_{dev},
   G_{vh} = (G_{vh} max)/25   in kPa

Masing rules applied for loading, unloading and reloading (see Figure 2.33).

Sources: Dasari 1996

Model 4   Strain-hardening plastic isotropic model

Cu, v, K as per Model 0 for undrained case

E = 1500*Cu   in kPa
Cohesion = 0.01*Cu   in kPa
Which initially is increased to $(\sigma_v\text{-}\sigma_h)/2$ since $K_0 \neq 1.0$
Cohesion = Cu*(1.0 – A/B)   for strain-hardening relationship   in kPa
A = 0.99*(1.0 – $((\sigma_v - \sigma_h)/2)/Cu$)
B = 1.0 + (190.0* ε_{dev})^2 + (145.0*ε_{dev})$^{0.56}$

The plasticity model is applied to a region from the surface to 22 m below tunnel axis level, within 22 m of the centreline in the X direction and for 64 m in the Y direction from the start of the tunnel (see Figure 6.3). Outside this area, the model is linear elastic, with a stiffness, E = 1500*Cu in kPa.

Sources: Pound 1999

Model 5   Linear elastic perfectly plastic isotropic model (high stiffness)

As per Model 4, except that in the plastic region, the yield criterion is based on Cu and the pre-yield behaviour is linear elastic, with E = 1500*Cu.
   * The cross-references refer to sections in the original thesis.

Appendix F

Thermo-chemo-mechanical constitutive model for sprayed concrete

The chemical affinity is defined as:

$$\frac{d\xi}{dt} = \left(1-\xi_0\right)\cdot\left[\frac{df_c/dt}{f_{c,\infty}}\right] = A_{(\xi)}\cdot e^{-E_A/RT_t} \qquad \text{(AF.1)}$$

$$A_{(\xi)} = a_A \cdot \left[\frac{1-e^{-b_A\cdot\xi}}{1+c_A\cdot\zeta^{d_A}}\right] \qquad \text{(AF.2)}$$

where the constants are

ξ_0 is the "percolation threshold" = 0.05 or calculated from Equation 5.23

a_A = 5.98
b_A = 18.02
c_A = 85.89
d_A = 7.377
E_A/R = 4200 where R is the universal gas constant and E_A is the activation energy of the cement

And f_c is the compressive strength and t is time. The figure below shows the growth of the normalised affinity predicted by equation 5.21 vs degree of hydration along with measured values.

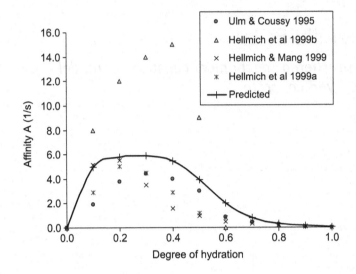

Appendix G

Frequency of testing from Austrian guidelines*

Type of test	Test	Minimum frequency according to control class			Source
		I	II	III	
Early age strength	30 needle or bolt test at early age from 15 min to 24 hours	1/mth**	1/mth	1/500 m²	ÖBV
Compressive strength	30 drilled cores at ages 7 & 28 days (generally)	1/5000 m² & 1/mth	1/2500 m² & 1/mth	1/250 m² & 2/mth	ÖBV
Water impermeability	ÖNORM B 3303 at 28 days	–	–	every 2 mth	ÖBV
Frost resistance	ÖNORM B 3303 at 56 days	–	1	5000 m² & every 3 mth	ÖBV
Sulphate resistance	ÖBV at 28 days	–	1	10000 m² & every 6 mth	ÖBV
Elastic modulus	ÖNORM B 3303 at 56 days	–	1	10000 m² & every 6 mth	ÖBV
Aggregate	Grading curve	1/mth	2/mth	4/mth	ÖBV
Aggregate	Moisture level	–	on-going	on-going	ÖBV
Cement	Specific surface area, setting, strength, water demand	–	every 2 mth	1/mth	ÖBV
Special cements	(according to the special requirement)				ÖBV
Additives (eg: PFA)	Specific surface area	–			ÖBV
Accelerator	Identification tests, lack of effect on reinforcement, etc.	every 2 mth	every 2 mth	1/mth	ÖBV
Dry spray mix	Composition	1/mth	2/mth	4/mth	ÖBV
Dry spray mix	Water content	1/mth	2/mth	4/mth	ÖBV

(Continued)

| Type of test | Test | Minimum frequency according to control class | | | Source |
		I	II	III	
Base concrete	Compressive strength and strength reduction	I	every 2 mth	every 2 mth	ÖBV

* This is an abbreviated reproduction of Tables 11/1 & 11/2, ÖBV Guideline on sprayed concrete (1998).
** mth = month.

Bibliography

ACI 209R (1992) "Prediction of Creep, Shrinkage and Temperature Effects in Concrete Structures", American Concrete Institute.

ACI 318-14 (2014) "Building Code Requirements for Structural Concrete and Commentary", American Concrete Institute.

ACI 506R-16 (2016) "Guide to Shotcrete", American Concrete Institute.

ACI 506.2 (2013) "Specification for Shotcrete", American Concrete Institute.

ACI-C660 (2002) "Certification for Shotcrete Nozzlemen", American Concrete Institute.

ACI 544.5R-10 (2010) "Report on the Physical Properties and Durability of Fiber-Reinforced Concrete", American Concrete Institute.

ASTM C403M-06 (2006) "Standard Test Method for Time of Setting of Concrete Mixtures by Penetration Resistance", ASTM International.

Abler, P. (1992) "Einflusse auf das Verformungsverhalten von jungem Spritzbeton im Tunnelbau", Diplomarbeit, University of Innsbruck.

Abu-Krisha, A.A.M. (1998) "Numerical modelling of TBM tunnelling in consolidated clay", PhD Thesis, University of Innsbruck.

Addenbrooke, T.I. (1996) "Numerical analysis of tunnelling in stiff clay", PhD Thesis, Imperial College London.

AFTES (2013) "Design, dimensioning and execution of precast steel fibre reinforced concrete arch segments", recommendations of AFTES No. GT38R1A1, *Tunnels et espaces souterrain*, No. 238, July/August 2013.

Aggistalis, G., Georganopoulos, C., Kazilis, N. & Game, R.C. (2004) "Problems and solutions in squeezing ground: Cases from Egnatia motorway tunnels", *Shotcrete: More Engineering Developments*, Bernard, E.S. (ed.), A. A. Balkema, Leiden, pp. 1–13.

Aggoun, S. & Torrenti, J.-M. (1996) "Effect of heat on the mechanical behaviour of concrete linings for tunnel", *Concrete for Infrastructure and Utilities*, Dhir & Henderson (eds.), E & F N Spon, London, pp. 161–170.

Aldrian, W. (1991) "Beitrag zum Materialverhalten von fruh belastetem Spritzbeton", Diplomarbeit, Montanuniversitat Leoben.

Aldrian, W. & Kattinger, A. (1997) "Monitoring of performance of primary support of NATM station at Heathrow Terminal 4", *Tunnels for People*, Hinkel, Golser & Schubert (eds.), Taylor & Francis, London, pp. 71–77.

Alkhiami, H. (1995) "Ein Naherungsverfahren zur Abschatzung der Belastung einer Spritzbetonkalottenschale auf der Grundlage von in-situ-Messungen", PhD Thesis, Hannover University.

Annett, M., Earnshaw, G. & Leggett, M. (1997) "Permanent sprayed concrete tunnel linings at Heathrow Airport", *Tunnelling '97*, IMM, pp. 517–534.

Ansell, A. (2004) "A finite element model for dynamic analysis of shotcrete on rock subjected to blast induced vibrations", *Shotcrete: More Engineering Developments*, Bernard, E.S. (ed.), A. A. Balkema, Leiden, pp. 15–25.

Ansell, A. (2011) "Shrinkage cracking in sprayed concrete on soft drains in traffic tunnels", *6th International Symposium on Sprayed Concrete*, Tromso, pp. 27–38.

Ansell, A., Bryne, L.E. & Holmgren, J. (2014) "Testing and evaluation of shrinkage cracking in sprayed concrete on soft drains", *7th International Symposium on Sprayed Concrete*, Sandefjord, pp. 20–32.

Archibald, J.F. & Dirige, P.A. (2006) "Development of thin, spray-on liner and composite superliner area supports for damage mitigation in blast- and rockburst-induced rock failure events", *Structures Under Shock and Impact IX, WIT Transactions on The Built Environment*, Vol. 87, WIT Press, pp. 237–246.

Arnold, J. & Neumann, C. (1995) "Umsetzung eines innovativen NOT-Konzeptes im Zuge eines "Know-how-Transfers"", *Felsbau*, Vol. 13, No. 6, pp. 459–463.

Asche, H.R. & Bernard, E.S. (2004) "Shotcrete design and specification for the Cross City Tunnel, Sydney", *Shotcrete: More Engineering Developments*, Bernard (ed.), A. A. Balkema, Leiden, pp. 27–38.

Atzwanger, R. (1999) "Die Sulfatbestandigkeit alkalifrei beschleunigter Spritzbetone", Diplomarbeit, University of Innsbruck.

Audsley, R.C., Favaloro, G. & Powell, D.B. (1999) "Design and implementation of the Heathrow Express Headshunt", *Tunnel Construction & Piling '99*, IMM, pp. 382–398.

Austin, S.A. & Robins, P.J. (1995) *Sprayed Concrete: Properties, Design and Application*, Whittles Publishing, Latheronwheel.

Austin, S.A., Robins, P.J. & Goodier, C.I. (2000) "Construction and repair with wet-process sprayed concrete and mortar", *The Concrete Society Technical Report 56*.

Austin, S.A., Robins, P.J. & Peaston, C.H. (1998) "Effects of silica fume on dry-process sprayed concrete", *Magazine of Concrete Research*, Vol. 58, No. 1, pp. 25–36.

Aydan, O., Sezaki, M. & Kawamoto, T. (1992a) "Mechanical and numerical modelling of shotcrete", *Numerical Models in Geomechanics*, Pande & Pietruszczak (eds.), Taylor & Francis, London, pp. 757–764.

Aydan, O., Sezaki, M., Kawata, T., Swoboda, G. & Moussa, A. (1992b) "Numerical modelling for the representation of shotcrete hardening and face advance of tunnels excavated by bench excavation method", *Numerical Models in Geomechanics*, Pande & Pietruszczak (eds.), Taylor & Francis, London, pp. 707–716.

BASF (2012) *Sprayed Concrete for Ground Support*, BASF Construction Chemicals Europe Ltd, Zurich.

BS8110 Part 1 (1997) "Structural use of concrete – Code of practice for design and construction", British Standards Institution, London.

BS8110 Part 2 (1985) "Structural use of concrete – Code of practice for special circumstances", British Standards Institution, London.

BTS (2004) *Tunnel Lining Design Guide*, British Tunnelling Society, Thomas Telford, London.

BTS (2010) *Specification for Tunnelling: Third Edition*, British Tunnelling Society, Thomas Telford, London.

BTS (2011) *Monitoring Underground Construction: A Best Practice Guide*, British Tunnelling Society, Thomas Telford, London.

BTS/ABI (2003) *Joint Code of Practice for Risk Management of Tunnel Works in the UK*, British Tunnelling Society / Association of British Insurers, London.

Banton, C., Diederichs, M.S., Hutchinson, D.J. & Espley, S. (2004) "Mechanisms of shotcrete roof support", *Shotcrete: More Engineering Developments*, Bernard (ed.), A. A. Balkema, Leiden, pp. 39–45.

Barratt, D.A., O'Reilly, M.P. & Temporal, J. (1994) "Long-term measurements of loads on tunnel linings in overconsolidated clay", *Tunnelling '94*, IMM, pp. 469–481.

Barrett, S.V.L. & McCreath, D.R. (1995) "Shotcrete support design in blocky ground: Towards a deterministic approach", *Tunnelling and Underground Space Technology*, Vol. 10, No. 1, pp. 79–89.

Barton, N. (2000) *TBM Tunnelling in Jointed and Faulted Rock*, Balkema, Rotterdam.

Barton, N., Lien, R. & Lunde J. (1975) "Estimation of support requirements for underground excavations", *ASCE Proceedings of 16th Symposium on Design Methods in Rock Mechanics*, Minnesota, MN, pp. 163–177.

Basso Trujilllo, P. & Jolin, M. (2018) "Encapsulation quality of reinforcement: From laboratory tests to structural design guidelines", *8th International Symposium on Sprayed Concrete*, Trondheim, pp. 54–68.

Beck, T. & Brusletto, K. (2018) "Sprayed concrete as avalanche securing of a road in the open day during freezing conditions", *8th International Symposium on Sprayed Concrete*, Trondheim, pp. 69–77.

Bernard, E.S. (2004a) "Creep of cracked fibre reinforced shotcrete panels", *Shotcrete: More Engineering Developments*, Bernard (ed.), A. A. Balkema, Leiden, pp. 47–57.

Bernard, E.S. (2004b) "Durability of cracked fibre reinforced shotcrete panels", *Shotcrete: More Engineering Developments*, Bernard (ed.), A. A. Balkema, Leiden, pp. 59–66.

Bernard, E.S. (2004c) "Design performance requirements for fibre-reinforced shotcrete using ASTM C-1550", *Shotcrete: More Engineering Developments*, Bernard (ed.), A. A. Balkema, Leiden, pp. 66–80.

Bernard, E.S. (2009) "Design of fibre reinforced shotcrete linings with macro-synthetic fibres", *ECI Conference on Shotcrete for Underground Support XI*, Amberg & Garshol (eds.), Davos, Switzerland, 2009.

Bernard, E.S. (2010) "Influence of fiber type on creep deformation of cracked fiber-reinforced shotcrete panels", *ACI Materials Journal*, Vol. 107, pp 474–480.

Bernard, E.S. (2014) "The use of macro-synthetic FRS for safe underground hard rock support", *Proceedings of the World Tunnel Congress 2014 – Tunnels for a better Life*, Foz do Iguaçu, Brazil.

Bernard, E.S. (2016) "Enhancement of seismic resistance of reinforced concrete members using embossed macro-synthetic fibers", *Proceedings of the World Tunnel Congress 2016*, San Francisco, USA.

Bernard, E.S. & Clements, M.J.K. (2001) "The influence of curing on the performance of fibre reinforced shotcrete panels", *Engineering Developments in Shotcrete*, Bernard, E.S. (ed.), Swets & Zeitlinger, Lisse, pp. 59–63.

Bernard, E.S., Clements, M.J.K., Duffield, S.B. & Morgan, D.R. (2014) "Development of macro-synthetic fibre reinforced shotcrete in Australia", *7th International Symposium on Sprayed Concrete*, Sandefjord, pp. 67–75.

Bertsch, A. (1992) "Qualitätsbeeinflussung von Spritzbeton durch Druckluftbeaufschlagung", Diplomarbeit, University of Innsbruck.

Bertuzzi, R. & Pells, P.J.N. (2002) "Design of rock bolt and shotcrete support of tunnel roofs in Sydney sandstone", *Australian Geomechanics*, Vol. 37, No. 3, pp. 81–90.

Berwanger, W. (1986) "Dreidimensionale Berechnung von tiefliegenden Felstunneln unter Berucksichtigung des rheologischen Verhaltens von Spritzbeton und des Bauverfahrens", Forschungsergebnisse aus dem Tunnel- und Kavernenbau, University of Hannover, Vol. 10.

Bezard, D. & Otten, G. (2018) "Metakaolin as a binder for free lime in sprayed concrete", *8th International Symposium on Sprayed Concrete*, Trondheim, pp. 78–83.

Bhewa, Y., Rougelot, T., Zghondi, J. & Burlion, N. (2018) "Instant and delayed mechanical behaviour of sprayed concrete used on Andra's URL", *8th International Symposium on Sprayed Concrete*, Trondheim, pp. 84–96.

Bieniawski, Z.T. (1984) *Rockmechanics Design in Mining and Tunneling*, Balkema, Rotterdam.

Bjontegaard, O. & Myren, S.A. (2011) "Fibre reinforced sprayed concrete panel tests: Main results from a methodology study performed by the Norwegian Sprayed Concrete Committee", *6th International Symposium on Sprayed Concrete*, Tromso, pp. 46–61.

Bjontegaard, O., Myren, S.A. & Beck, T. (2018) "Quality control of fibre reinforced sprayed concrete: Norwegian requirements and experiences from laboratory studies and tunnel projects", *8th International Symposium on Sprayed Concrete*, Trondheim, pp. 97–107.

Bjontegaard, O., Myren, S.A., Klemestrud, K., Kompen, R. & Beck, T. (2014) "Fibre reinforced sprayed concrete (FRSC): Energy absorption capacity from 2 days age to one year", *7th International Symposium on Sprayed Concrete*, Sandefjord, pp. 88–97.

Blasen, A. (1998) "Bestimmung von Porositätskennwerten am Spritzbeton und deren Einfluss auf betontechnologische Parameter", Diplomarbeit, University of Innsbruck.

Bloodworth, A. & Su, J. (2018) "Numerical analysis and capacity evaluation of composite sprayed concrete lined tunnels", *Underground Space*, Vol. 3, No. 2, June 2018, pp. 87–108.

Bolton, A. (1999) "Mont Blanc: The Aftermath", *New Civil Engineer*, 15 April 1999, pp. 18–20.

Bolton, A. & Jones, M. (1999) "Second road tunnel tragedy heightens safety fears", *New Civil Engineer*, 3 June 1999, p. 3.

Bolton, M.D., Dasari, G.R. & Rawlings, C.G. (1996) "Numerical modelling of a NATM tunnel construction in London Clay", *Geotechnical Aspects of Underground Construction in Soft Ground*, Mair & Taylor (eds.), Taylor & Francis, London, pp. 491–496.

Bonapace, P. (1997) "Evaluation of stress measurements in NATM tunnels at the Jubilee Line Extension Project", *Tunnels for People*, Hinkel, Golser & Schubert (eds.), Taylor & Francis, London, pp. 325–330.

Boniface, A. & Morgan, D.R. (2009) "Durability and curing of shotcrete'", *ITA/SAIMM/SANCOT "Shotcrete for Africa" Conference*, Johannesburg, 2–3 March 2009.

Bonin, K. (2012) "Abdichtung mit spritzbeton in einschaliger bauweise", Spritzbeton-Tagung, 2012.

Brite Euram C1 (1997) "Sub Task C1: Development of time-dependent mathematical model", BRE-CT92-0231, Institute of Mechanics of Materials and Geostructures.

Brite Euram C2 (1997) "Sub Task C2: Collapse limit-state model", BRE-CT92-0231, Imperial College London.

Brite Euram D1 (1997) "Sub-Task D1: Structural design guidelines", BRE-CT92-0231 (MM ref: 06460/D01/A), Mott MacDonald Ltd.

Brite Euram (1998) "New materials, design and construction techniques for underground structures in soft rock and clay media", BRE-CT92-0231 Final Technical Report, Mott MacDonald Ltd.

Brooks, J. (1999) "Shotcrete for ground support as used in the Asia Pacific region", *Rapid Excavation and Tunnelling Conference Proceedings 1999*, Orlando, FL, pp. 473–524.

Bryne, L.E., Ansell, A. & Holmgren, J. (2014) "Early age bond strength between hard rock and hardening sprayed concrete", *7th International Symposium on Sprayed Concrete*, Sandefjord, pp. 112–123.

Bryne, L.E., Holmgren, J. & Ansell, A. (2011) "Experimental investigation of the bond strength between rock and hardening sprayed concrete", *6th International Symposium on Sprayed Concrete*, Tromso, pp. 77–88.

Burd, H.J., Houlsby, G.T., Augarde, C.E. & Liu, G. (2000) "Modelling tunnelling-induced settlement of masonry buildings", *Proceedings Institution of Civil Engineers, Civil Engineering*, Vol. 143, January, pp. 17–29.

Burland, J.B., Standing, J.R., & Jardine, F.M. (2001) *Building Response to Tunnelling. Case Studies from the Jubilee Line Extension, London*, CIRIA / Thomas Telford, London.

Byfors, J. (1980) *Plain Concrete at Early Ages*, Research Report, Swedish Cement & Concrete Research Institute, Dept. of Building, Royal Institute of Technology, Stockholm.

Celestino, T.B. (2005) "Shotcrete and waterproofing for operational tunnels", *Proceedings of ITA Workshop on Waterproofing*, Sao Paulo, 2005, www.ita-aites.org.

Celestino, T.B., Bortolucci, A.A., Re, G. & Ferreira, A.A. (1999) "Diametral compression tests for the determination of shotcrete anisotropic elastic constants", *Shotcrete for Underground VIII*, Sao Paulo.

Cervera, M., Oliver, J. & Prato, T. (1999a) "Thermo-chemo-mechanical model for concrete. II: Hydration and aging", *ASCE Journal of Engineering Mechanics, ASCE*, Vol. 125, No. 9, pp. 1018–1027.

Cervera, M., Oliver, J. & Prato, T. (1999b) "Thermo-chemo-mechanical model for concrete. II: Damage and creep", *ASCE Journal of Engineering Mechanics, ASCE*, Vol. 125, No. 9, pp. 1028–1039.

Chang, Y. (1994) "Tunnel support with shotcrete in weak rock - a rock mechanics study", PhD Thesis, Royal Institute of Technology, Stockholm.

Chang, Y. & Stille, H. (1993) "Influence of early-age properties of shotcrete on tunnel construction sequences", *Shotcrete for Underground Support VI*, American Society of Civil Engineers, Reston, pp. 110–117.

Chapman, D.N., Nicole Metje, N. & Stark, A. (2017) *Introduction to Tunnel Construction*, CRC Press, Boca Raton, FL.

Chen, W.F. (1982) *Plasticity in Reinforced Concrete*, McGraw-Hill, Inc., New York.

Chen, W.-P.N. & Vincent, F. (2011) "Single-shell synthetic fiber shotcrete lining for SLAC tunnels", *Rapid Excavation and Tunneling Conference Proceedings 2011*, San Francisco, CA, pp. 1336–1355.

Choi, S., Thienel, K.-C. & Shah, S.P. (1996) "Strain softening of concrete in compression under different end conditions", *Magazine of Concrete Research*, Vol. 48, No. 175, pp. 103–115.

Clayton, C.R.I., Hope, V.S., Heyman, G., van der Berg, J.P. & Bica, A.V.D. (2000) "Instrumentation for monitoring sprayed concrete lined soft ground tunnels", *Proceedings Institution of Civil Engineers, Geotechnical Engineering*, Vol. 143, pp. 119–130.

Clayton, C.R.I., van der Berg, J.P., Heyman, G., Bica, A.V.D. & Hope, V.S. (2002) "The performance of pressure cells for sprayed concrete tunnel linings", *Geotechnique*, Vol. 52, No. 2, pp. 107–115.

Clayton, C.R.I., van der Berg, J.P. & Thomas, A.H. (2006) "Monitoring and displacements at Heathrow Express Terminal 4 station tunnels", *Geotechnique*, Vol. 56, No. 5, pp. 323–334.

Clements, M.J.K. (2004) "Comparison of methods for early age strength testing of shotcrete", *Shotcrete: More Engineering Developments*, Bernard (ed.), A. A. Balkema, Leiden, pp. 81–87.

Clements, M.J.K., Jenkins, P.A. & Malmgren, L. (2004) "Hydro-scaling – An overview of a young technology", *Shotcrete: More Engineering Developments*, Bernard (ed.), A. A. Balkema, Leiden, pp. 89–96.

CLRL (2014) "Best Practice Guide - SCL Exclusion Zone Management (Crossrail)", CRL1-XRL-C-GUI-CR001-50001, www.britishtunnelling.org.uk.

Collis, A. & Mueller, A.M.O. (2004) "Recommendations for future developments in shotcrete: A contractor's perspective", *Shotcrete: More Engineering Developments*, Bernard (ed.), A. A. Balkema, Leiden, pp. 97–101.

Concrete Society (1988) "Permeability testing of site concrete: A review of methods and experience", *Technical Report 31*, The Concrete Society.

Cornejo-Malm, G. (1995) "Schwinden von Spritzbeton", *Internal Report*, ETH Zurich.

Cosciotti, L., Lembo-Fazio, A., Boldini, D. & Graziani, A. (2001) "Simplified behavior models of tunnel faces supported by shotcrete and bolts", *Proceedings of the International Conference on Modern Tunneling Science and Technology (IS-Kyoto 2001)*, T. Adachi, K. Tateyama, & M. Kimura (eds.), Kyoto, Vol. 1, pp. 407–412.

Crehan, D., Eide, S. & Vlietstra, D. (2018) "Sustainability and the responsible disposal of contaminated waste", *8th International Symposium on Sprayed Concrete*, Trondheim, pp. 108–115.

Curtis, D.J. (1976) "Discussions on Muir-Wood, The circular tunnel in elastic ground", *Geotechnique*, Vol. 26, No. 1, pp. 231–237.

DBV (1992) "Design basis for steel fibre reinforced concrete in tunnel construction", Code of Practice, Deutscher Beton Verein.

DIN 1045 (1988) "Structural use of concrete. Design and construction", Deutsches Institut für Normung, e.V..

DIN 1048 (1991) "Testing concrete", Deutsches Institut für Normung, e.V.

DIN 18551 (1992) "Sprayed concrete: Production and inspection", Deutsches Institut für Normung, e.V.

D'Aloia & Clement (1999) "Meeting compressive strength requirements at early age by using numerical tools - determination of apparent activation energy of concrete", *Modern Concrete Materials: Binders, Additions and Admixtures*, Dhir & Dyer (eds.), Thomas Telford, London, pp. 637–652.

Darby, A. & Leggett, M. (1997) "Use of shotcrete as the permanent lining of tunnels in soft ground", Mott MacDonald Milne Award submission (unpublished).

Dasari, G.R. (1996) "Modelling the variation of soil stiffness during sequential construction", PhD Thesis, University of Cambridge.

Davik, K. & Markey, I. (1997) "Durability of sprayed concrete in Norwegian road tunnels", *Tunnelling '97*, IMM, pp. 251–261.

Deane, A.P. & Bassett, R.H. (1995) "The Heathrow Express trial tunnel", *Proceedings Institution of Civil Engineers, Geotechnical Engineering*, Vol. 113, pp. 144–156.

De Battista, N., Elshafie, M., Soga, K., Williamson, M., Hazelden, G. & Hsu, Y.S. (2015) "Strain monitoring using embedded distributed fibre optic sensors in a sprayed concrete tunnel lining during the excavation of cross-passages", *Proceedings of the 7th International Conference on Structural Health Monitoring of Intelligent Infrastructure*, International Society for Structural Health Monitoring of Intelligent Infrastructure, Winnipeg.

Denney, J.M. & Hagan, P.C. (2004) "A study on the effect of changes in fibre type and dosage rate on Fibre Reinforced Shotcrete performance", *Shotcrete: More Engineering Developments*, Bernard (ed.), A. A. Balkema, Leiden, pp. 103–108.

Diederichs, M.S. & Kaiser P.K. (1999) "Stability guidelines for excavations in laminated ground - the voussoir analogue revisited", *International Journal of Rock Mechanics and Mining Sciences*, Vol. 36, pp. 97–118.

Dimmock, R. (2011) "SCL in the UK", *Tunnelling Journal*, December 2010/ January 2011, pp. 20–23.

Ding, Y. (1998) "Technologische Eigenschaften von jungem Stahlfaserbeton und Stahlfaserspritzbeton", PhD Thesis, University of Innsbruck.

DiNoia, T.P. & Rieder, K.-A. (2004) "Toughness of fibre-reinforced shotcrete as a function of time, strength development and fiber type according to ASTM C1550-02", *Shotcrete: More Engineering Developments*, Bernard (ed.), A. A. Balkema, Leiden, pp. 127–135.

DiNoia, T.P. & Sandberg, P.J. (2004) "Alkali-free shotcrete accelerator interactions with cement and admixtures", *Shotcrete: More Engineering Developments*, Bernard (ed.), A. A. Balkema, Leiden, pp. 137–144.

EFNARC (1996) "European Specification for Sprayed Concrete" , European Federation of Producers and Applicators of Specialist Products for Structures. www.efnarc.org

EFNARC (1999) "Execution of spraying", European Federation of Producers and Applicators of Specialist Products for Structures. www.efnarc.org.

EFNARC (2002) "European specification for sprayed concrete", European Federation of Producers and Applicators of Specialist Products for Structures. www.efnarc.org.

EN 14487 (2006) "Sprayed concrete", British Standards Institution, London.

EN 14488 (2006) "Testing Sprayed Concrete", British Standards Institution, London.

EN14889-1 (2006) "Fibres for concrete Part 1: Steel fibres - Definitions, specifications and conformity", British Standards Institution, London.

EN14889-2 (2006) "Fibres for concrete Part 2: Polymer fibres – Definitions, specifications and conformity", British Standards Institution, London.

EN206-1 (2000) "Concrete. Specification, Performance, Production and Conformity", British Standards Institution, London.

Eberhardsteiner, J., Meschke, G. & Mang, H.A. (1987) "Comparison of constitutive models for triaxially loaded concrete", *Proceedings of the IABSE Colloquium on Computational Mechanics of Concrete Structures in Delft*, IABSE, Zurich, pp. 197–208.

Eierle, B. & Schikora, K. (1999) "Computational modelling of concrete at early ages", *Diana World*, Vol. 2, p. 99.

Eichler, K. (1994) "Umweltfreundlicher Spritzbeton: Environmentally friendly shotcrete", *Tunnel*, Vol. 1, No. 94, pp. 33–37.

Einstein, H.H. & Schwartz, C.W. (1979) "Simplified analysis for tunnel supports", *ASCE Journal of Geotechnical Engineers Division*, Vol. 105 GT4, pp. 499–518.

Ellison, T.P., Franzen, T. & Karlsson, B.I. (2002) "Shotcrete use in Southern Link tunnel and some current shotcrete research in Sweden", *Spritzbeton Technologie 2002*, Alpbach, pp. 57–65.

England, G.L. & Illston, J.M. (1965) "Methods of computing stress in concrete from a history of measured strain parts 1, 2 & 3", *Civil Engineering and Public Works Review*, Issues April (pp. 513–517), May (pp. 692–694) & June (pp. 846–847).

Eurocode 2 (2004) EN 1992-1-1:2004 "Design of concrete structures — Part 1-1: General rules and rules for buildings", British Standards Institution, London.

Eurocode 7 (2004) EN1997-1:2004 "Geotechnical design — Part 1: General rules", British Standards Institution, London.

Everton, S. (1998) "Under observation", *Ground Engineering*, May, pp. 26–29.

Feenstra, P.H. & de Borst, R. (1993) "Aspects of robust computational models for plain and reinforced concrete", *Heron*, Vol. 48, No. 4, pp. 5–73.

fib (2010) "Model Code 2010", CEB-FIP, Lausanne, Switzerland.

Fischer, M. & Hofmann, M. (2015) "Reinforced shotcrete with bar diameters up to 32 mm", *Crossrail Learning Legacy*, https://learninglegacy.crossrail.co.uk/.

Fischnaller, G. (1992) "Untersuchungen zum Verformungsverhalten von jungem Sprtizbeton im Tunnelbau: Grundlagen und Versuche", Diplomarbeit, University of Innsbruck.

Franzen, T. (2005) "Waterproofing of tunnels in rock", *Proceedings of ITA Workshop on Waterproofing*, Sao Paulo 2005, www.ita-aites.org.

Franzen, T., Garshol, K.F. & Tomisawa, N. (2001) "Sprayed concrete for final linings: ITA working group report", *Tunnelling and Underground Space Technology*, Vol. 16, pp. 295–309.

Galan, I., Thumann, M., Briendl, L., Rock, R., Steindl, F., Juhart, J., Mittermayr, F. & Kusterle, W. (2018) "From lab scale spraying to real scale shotcreting and back to the lab", *8th International Symposium on Sprayed Concrete*, Trondheim, pp. 129–143.

Galobardes, I., Cavalaro, S.H., Aguado, A., Garcia, T. & Rueda, A. (2014) "Correlation between the compressive strength and the modulus of elasticity of sprayed concrete", *7th International Symposium on Sprayed Concrete*, Sandefjord, pp. 148–160.

Garshol, K.F. (2002) "Admixtures and other factors influencing durability of sprayed concrete", *4th International Symposium on Sprayed Concrete*, Sandefjord, pp. 123–129.

Gebauer, B. (1990) "The single permanent shotcrete lining method for the construction of galleries and traffic tunnels – the result of practice-oriented research and application", *Proceedings of the 3rd Conference on Shotcrete Technology*, Innsbruck-Igls: Institut für Baustofflehre und Materialprüfung, Universität Innsbruck, pp. 41–58.

Geoguide 4 (1992) *"Guide to Cavern Engineering"*, Geotechnical Engineering Office, Civil Engineering Department, Hong Kong.

Gerstle, K.H. (1981) "Simple formulation of biaxial concrete behaviour", *American Concrete Institute Journal*, Vol. 78, No. 1, pp. 62–68.

Gibson, A. & Bernard, E.S. (2011) "The early-age strength evaluation for FRS using embedded UPV measurement", *6th International Symposium on Sprayed Concrete*, Tromso, pp. 151–160.

Goit, C.S., Kovács, A. & Thomas, A.H. (2011) "Advanced numerical modelling in tunnel design – the example of a major project in the UK", *2nd International FLAC/DEM Symposium*, 14–16 February 2011, Melbourne, Australia, pp. 117–126.

Golser, J. (1999) "Behaviour of early-age shotcrete", *Shotcrete for Underground VIII*, Sao Paulo.

Golser, J. & Kienberger, G. (1997) "Permanente Tunnelauskleidung in Spritzbeton - Beanspruchung und Sicherheitsfragen", *Felsbau*, No. 6, pp. 416–421.

Golser, J., Schubert, P. & Rabensteiner, K. (1989) "A new concept for evaluation of loading in shotcrete linings", *Proceedings of International Congress on Progress and Innovation in Tunnelling*, Lo, K.Y. (ed.), Tunnelling Association of Canada, Toronto, pp. 79–85.

Goransson, E., Loncaric, A.J. & Singh, U. (2014)"Robotic shotcrete operator training in Australia", *7th International Symposium on Sprayed Concrete*, Sandefjord, pp. 185–197.

Greman, A. (2000) "AlpTransit's first major excavations", *Tunnels & Tunnelling International*, October 2000, pp. 22–35.

Grimstad, E. (1999) "Experiences from excavation under high rock stress in the 24,5 km long Laerdal Tunnel", *International Congress on Rock Engineering Techniques for Site Characterisation*, Bangalore, India, pp. 135–146.

Grimstad, E. & Barton, N. (1993) "Updating the Q-system for NMT", *Proceedings of the International Symposium on Sprayed Concrete*, Norwegian Concrete Association, Oslo, pp. 46–66.

Grose, W.J. & Eddie, C.M. (1996) "Geotechnical aspects of the construction of the Heathrow Transfer Baggage System tunnel", *Geotechnical Aspects of Underground Construction in Soft Ground*, Mair & Taylor (eds.), Taylor & Francis, London, pp. 269–276.

Grov, E. (2011) "Sprayed concrete as an integrated part of Norwegian tunnelling including some international examples", *6th International Symposium on Sprayed Concrete*, Tromso, pp. 8–26.

Guilloux, A., le Bissonnais, H., Robert, J. & Bernardet, A. (1998) "Influence of the K0 coefficient on the design of tunnels in hard soils", *Tunnels and Metropolises*, Negro Jr & Ferreira (eds.), Taylor & Francis, London, pp. 387–392.

Gunn, M.J. (1993) "The prediction of surface settlement profiles due to tunnelling", *Predictive Soil Mechanics*, Thomas Telford, London.

HA (2006) "Volume 1 Specification for Highway Works", Highways Agency, Manual of Contract Documents for Highway Works, HMSO, Norwich, UK.

HSE (1996) *Safety of New Austrian Tunnelling Method (NATM) Tunnels*, Health & Safety Executive, HMSO, Norwich, UK.

HSE (2000) *The Collapse of NATM Tunnels at Heathrow Airport*, HSE Books, HMSO, Norwich, UK.

Hafez, N.M. (1995) "Post-failure modelling of three-dimensional shotcrete lining for tunnelling", PhD thesis, University of Innsbruck.

Hagelia, P. (2018) "Durability of sprayed concrete for rock support: A tale from the tunnels", *8th International Symposium on Sprayed Concrete*, Trondheim, pp. 172–187.

Han, N. (1995) "Creep of high strength concrete", *Progress in Concrete Research*, Vol. 4, pp. 107–118.

Hannant, D.J., Branch, J. & Mulheron, M. (1999) "Equipment for tensile testing of fresh concrete", *Magazine of Concrete Research*, Vol. 51, No. 4 August, pp. 263–267.

Hashash, Y.M.A., Hook, J.J., Schmidt, B. & Yao, J.I.-C. (2001) "Seismic design and analysis of underground structures", *Tunnelling and Underground Space Technology*, No. 16, 2001, pp. 247–293.

Hauck, C. (2014) "Field experiences from two Large Tunnel Projects in Norway", *7th International Symposium on Sprayed Concrete*, Sandefjord, pp. 198–207.

Hauck, C., Bathen, L. & Mathisen, A.E. (2011) "The effect of air entraining admixture on sprayed concrete fresh- and hardened properties", *6th International Symposium on Sprayed Concrete*, Tromso, pp. 190–202.

Hauck, C., Mathisen, A.E. & Grimstad, E. (2004) "Macro-synthetic fibre reinforced shotcrete in a Norwegian road tunnel", *Shotcrete: More Engineering Developments*, Bernard (ed.), A. A. Balkema, Leiden, pp. 161–168.

Haugeneder, E., Mang, H., Chen, Z.S., Heinrich, R., Hofstetter, G., Li, Z.K., Mehl, M. & Torzicky, P., (1990) "3D Berechnungen von Tunnelschalen aus Stahlbeton", *Strassenforschung Heft 382*, Vienna.

Hawley, J. & Pottler, R. (1991) "The Channel Tunnel: Numerical models used for design of the United Kingdom undersea crossover", *Tunnelling 91*, IMM, 1991.

Hefti, R. (1988) "Einfluss der Nachbehandlung auf die Spritzbetonqualität", Research report from ETH Zurich.

Hellmich, C. & Mang, H.A. (1999) "Influence of the dilatation of soil and shotcrete on the load bearing behaviour of NATM-tunnels", *Felsbau*, Vol. 17, No. 1, pp. 35–43.

Hellmich, C., Mang, H.A., Schon, E. & Friedle, R. (1999a) "Materialmodellierung von Spritzbeton – vom Experiment zum konstitutiven Gesetz", Report of internal seminar, TU Wien.

Hellmich, C., Ulm, F.-J. & Mang, H.A. (1999b) "Multisurface chemoplasticity. I: Material model for shotcrete", *ASCE Journal of Engineering Mechanics*, Vol. 125, No. 6, pp. 692–701.

Hellmich, C., Ulm, F.-J. & Mang, H.A. (1999c) "Multisurface chemoplasticity. II: Numerical studies on NATM tunneling", *ASCE Journal of Engineering Mechanics*, Vol. 125, No. 6, pp. 702–713.

Hellmich, C., Sercombe, J., Ulm, F.-J. & Mang, H. (2000) "Modeling of early-age creep of shotcrete. II: Application to Tunneling", *ASCE Journal of Engineering Mechanics*, Vol. 126, No. 3, pp. 292–299.

Hendron, A.J., Jr & Fernandez, G. (1983) "Dynamic and static design considerations for underground chambers", *Seismic Design of Embankments and Caverns*, ASCE, New York, pp. 157–197.

Henke, A. & Fabbri, D. (2004) "The Gotthard base tunnel: Project overview", *Proceedings of 7th International Symposium on Tunnel Construction and Underground Structures*, DzPiGK / University of Ljubljana, Ljubljana, Slovenia, pp. 107–116.

Hilar, M. & Thomas, A.H. (2005) "Tunnels construction under the Heathrow airport", *Tunnel*, Vol. 3, 2005, pp. 17–23.

Hilar, M., Thomas, A.H. & Falkner, L. (2005) "The latest innovation in sprayed concrete lining – the Lasershell Method", *Tunnel*, Vol. 4, 2005, pp. 11–19.

Hirschbock, U. (1997) "2D FE Untersuchungen zur Neuen Österreichischen Tunnelbaumethode", Diplomarbeit, TU Wien.

Hoek, E. & Brown, E.T. (1980) *Underground Excavations in Rock*, Institution of Mining and Metallurgy, London.

Hoek, E., Kaiser, P.K. & Bawden, W.F. (1998) *Support of Underground Excavations in Hard Rock*, Balkema, Rotterdam.

Hoek, E. & Marinos, P. (2000) "Deformation: Estimating rock mass strength", *Tunnels & Tunnelling International*, November 2000, pp. 45–51.

Hofstetter, G., Oettl, G. & Stark, R. (1999) "Development of a three-phase soil model for the simulation of tunnelling under compressed air", *Felsbau*, Vol. 17, No. 1, pp. 26–31.

Holmgren, J.B. (2004) "Experiences from shotcrete works in Swedish hard rock tunnels", *Shotcrete: More Engineering Developments*, Bernard (ed.), A. A. Balkema, Leiden, pp. 169–173.

Holter, K.-G. (2015a) "Properties of waterproof sprayed concrete tunnel linings. A study of EVA-based sprayed membranes for waterproofing of rail and road tunnels in hard rock and cold climate", PhD Thesis, Norwegian University of Science & Technology.

Holter, K.-G. (2015b) "Performance of EVA-based membranes for SCL in hard rock", *Rock Mechanics and Rock Engineering*, April 2016, Vol. 49, No. 4, pp. 1329–1358.

Holter, K.-G. & Geving, S. (2016) "Moisture transport through sprayed concrete tunnel linings", *Rock Mechanics and Rock Engineering*, January 2016, Vol. 49, No. 1, pp. 243–272.

Hrstka, O., Cerny, R. & Rovnanikova, P. (1999) "Hygrothermal stress induced problems in large scale sprayed concrete structures", *Specialist Techniques & Materials for Concrete Construction*, Dhir & Henderson (eds.), Thomas Telford, London, pp. 103–109.

Huang, Z. (1991) "Beanspruchungen des Tunnelbaus bei zeitabhängigem Materialverhalten von Beton und Gebirge", Inst. Für Statik Report No. 91 - 68, TU Braunschweig.

Huber, H.G. (1991) "Untersuchungen zum Verformungsverhalten von jungem Spritzbeton im Tunnelbau", Diplomarbeit, University of Innsbruck.

Hughes, T.G. (1996) "Flat Jack Investigation of Heathrow Terminal 4 Concourse Tunnel", Report for the University of Surrey by University of Wales, Cardiff.

ICE (1996) *Sprayed Concrete Linings (NATM) for Tunnels in Soft Ground*, Institution of Civil Engineers design and practice guides, Thomas Telford, London.

ITA (1991) "Water leakages in subsurface facilities: Required watertightness, contractual methods and methods of redevelopment", *Tunnelling and Underground Space Technology*, Hauck, A. (ed.), Vol. 6, No. 3, pp. 273–282.

ITA (1993) "Shotcrete for rock support: A summary report on the state of the art in 15 countries", Malmberg, B. (ed.), *Tunnelling and Underground Space Technology*, Vol. 8, No. 4, pp. 441–470.

ITA (2004) "Guidelines for Structural Fire Resistance for Road Tunnels", Report of Working Group 6 Maintenance and Repair, www.ita-aites.org.

ITA (2006) "Shotcrete for rock support: A summary report on state-of-the-art", Report of Working Group 12 – Shotcrete Use, www.ita-aites.org.

ITA (2008) "Guidelines for good occupational health and safety practice in tunnel construction", Report of Working Group 5 – Health and Safety in Works, www.ita-aites.org.

ITA (2010) "Shotcrete for rock support: A summary report on state-of-the-art", Report of Working Group 12 – Sprayed Concrete Use, www.ita-aites.org.

ITA (2017) "Structural fire protection for road tunnels", Report of Working Group 6 – Maintenance and Repair, www.ita-aites.org.

ITAtech (2013) "Design guideline for spray applied waterproofing membranes", ITAtech Report No. 2 – April 2013, ISBN: 978-2-9700858-1-2.

ITAtech (2016) "ITAtech guidance for precast fibre reinforced concrete segments – vol. 1: Design aspects", ITAtech Report No. 7 – April 2016, ISBN: 978-2-9701013-2-1.

ITC (2006) "Mucking Tunnel Taglesberg", ITC SA News 30, 2006, http://www.itcsa.com/images/Stories/pdf/News_30_Taglesberg_AN.pdf.

Jager, J. (2016) "Structural design of composite shell linings", Proceedings of the World Tunnel Congress 2016, San Francisco, CA.

Jaeger, J.C. & Cook, N.G.W. (1979) Fundamentals of Rock Mechanics, Chapman and Hall, London.

Jahn, M. (2011) "Sprayed concrete application directly onto polymer waterproofing membranes: Example of Lungern Bypass (Switzerland)", 6th International Symposium on Sprayed Concrete, Tromso, pp. 203–211.

John, M. (1978) "Design of the Arlberg expressway tunnel and the Pfandertunnel", Shotcrete For Underground Support III, ASCE, St. Anton am Arlberg, Austria, 1978, pp. 27–43.

John, M. & Mattle, B. (2003) "Factors of shotcrete lining design", Tunnels & Tunnelling International, October 2003, pp. 42–44.

Jolin, M., Bolduc, L.-S., Bissonnette, B. & Power, P. (2011) "Long-term durability of sprayed concrete", 6th International Symposium on Sprayed Concrete, Tromso, pp. 212–225.

Jolin, M., Gagnon, F. & Beaupré, D. (2004) "Determination of criteria for the acceptance of shotcrete for certification", Shotcrete: More Engineering Developments, Bernard (ed.), A. A. Balkema, Leiden, pp. 175–181.

Jones, B.D. (2005) "Measurements of ground pressure on sprayed concrete tunnel linngs using radial pressure cells", Underground Construction 2005, Brintex, London.

Jones, B.D. (2007) "Stresses in sprayed concrete tunnel junctions", EngD Thesis, University of Southampton.

Jones, B.D. (2018) "A 20 year history of stress and strain in a shotcrete primary lining", 8th International Symposium on Sprayed Concrete, Trondheim, pp. 206–222.

Jones, B.D., Li, S. & Ahuja, V. (2014) "Early strength monitoring of shotcrete using thermal imaging", 7th International Symposium on Sprayed Concrete, Sandefjord, pp. 245–254.

Jones, B.D., Staerk, A. & Thomas, A.H. (2005) "The importance of stress measurement in a holistic sprayed concrete tunnel design process", Conference on Tunnelling for a sustainable Europe in Chambery, L'Association Française des Travaux en Souterrain (AFTES), Paris, pp. 433–440.

Jung, H.-I., Pillai, A., Wilson, C., Clement, F. & Traldi, D. (2017a) "Sprayed concrete composite shell lining – Part 1", Tunnelling Journal, September 2017, pp. 18–27.

Jung, H.-I., Pillai, A., Wilson, C., Clement, F. & Traldi, D. (2017b) "Sprayed concrete composite shell lining – Part 2", Tunnelling Journal, October/November 2017, pp. 33–39.

Kaiser, P.K. & Cai, M. (2012) "Design of rock support system under rockburst condition", Journal of Rock Mechanics and Geotechnical Engineering. 2012, Vol. 4, No. 3, pp. 215–227.

Kammerer, G. & Semprich, S. (1999) "The prediction of the air loss in tunnelling under compressed air", *Felsbau*, Vol. 17, No. 1, pp. 32–35.

Kaufmann, J., Bader, R. & Manser, M. (2012) "Untersuchungen zum Biege-Kriechverhalten von Faserbeton mit Makro-Synthetischen Bikomponentenfasern", *Spritzbeton-Tagung*, 2012.

Kaufmann, J. & Frech, K. (2011) "Rebound of plastic fibers in sprayed concrete applications", *6th International Symposium on Sprayed Concrete*, Tromso, pp. 232–241.

Kaufmann, J. & Manser, M. (2013) "Durability performance of bi-component polymer fibres under creep and in aggressive environments", *Ground Support 2013*, Brady & Potvin (eds.), Perth, Australia, pp. 585–596.

Kaufmann, J., Loser, R., Winnefeld, F. & Leemann, A. (2018) "Sulfate resistance and phase composition of modern shotcrete", *World Tunnel Congress 2018*, Dubai.

Kavanagh, T. & Haig, B. (2012) "Small scale SCL projects: Trials and Tribulations", *Tunnelling Journal*, February/March 2012, pp. 30–32.

Kimmance, J.P. & Allen, R. (1996) "NATM and compensation grouting trial at Redcross Way", *Geotechnical Aspects of Underground Construction in Soft Ground*, Mair & Taylor (eds.), Taylor & Francis, London, pp. 385–390.

King, M., St. John, A., Brown, D. & Comins, J. (2016) "Sprayed concrete lining falls and exclusion zone management", https://learninglegacy.crossrail.co.uk/documents/sprayed-concrete-lining-falls-exclusion-zone-management/.

Klados, G. (2002) "Sprayed concrete lining as temporary support in frozen soil", *4th International Symposium on Sprayed Concrete*, Davos, 2002, pp. 198–207.

Kodymova, J., Thomas, A.H. & Will, M. (2017) "Life-cycle assessments of rock bolts", *Tunnelling Journal*, June/July 2017, pp. 47–49.

Kompen, R. (1990) "Wet process steel fibre reinforced shotcrete for rock support and fire protection, Norwegian practice and experiences", *Spritzbeton Technologie '90*, pp. 87–92.

Kotsovos, M.D. & Newman, J.B. (1978) "Generalized stress-strain relations for concrete", *ASCE Journal of Engineering Mechanics Division*, Vol. 104, EM4, pp. 845–856.

Kovari, K. (1994) "On the existence of the NATM: Erroneous concepts behind the New Austrian Tunnelling Method", *Tunnels & Tunnelling*, November, pp. 38–42.

Krenn, F. (1999) "Small strain stiffness and its influence on the pattern of ground behaviour around a tunnel", Diplomarbeit, TU Graz.

Kropik, C. (1994) "Three-dimensional elasto-viscoplastic finite element analysis of deformations and stresses resulting from the excavation of shallow tunnels", PhD Thesis, TU Wien.

Kullaa, J. (1997) "Finite element modelling fibre-reinforced brittle materials", *Heron*, Vol. 42, No. 2, pp. 75–95.

Kupfer, H. & Kupfer, H. (1990) "Statical behaviour and bond performance of the layers of a single permanent shotcrete tunnel lining", *Proceedings 3rd Conference on Shotcrete Technology*, Innsbruck-Igls: Institut für Baustofflehre und Materialprufung, Universität Innsbruck, pp. 11–18.

Kusterle, W. (1992) "Qualitatsverbesserungen beim Spritzbeton durch technologische Massnahmen, durch den Einsatz neuer Materialien und auf Grund der Erfassung von Spritzbetoneigenschaften", Habilationsschrift, Vol. 1 & 2, University of Innsbruck.

Kuwajima, F.M. (1999) "Early age properties of the shotcrete", *Shotcrete for Underground VIII*, Sao Paulo.

Lackner, R. (1995) "Ein anisotropes Werkstoffmodell für Beton auf der Grundlage der Plastitatstheorie und der Schadigungstheorie", Diplomarbeit, TU Wien.

Lamis, A. (2018) "Vulnerability of shotcrete on tunnel walls during construction blasting", *World Tunnel Congress 2018*, Dubai.

Lamis, A. & Ansell, A. (2014) "Behaviour of sprayed concrete on hard rock exposed to vibrations from blasting operations", *7th International Symposium on Sprayed Concrete*, Sandefjord, pp. 7–19.

Laplante, P. & Boulay, C. (1994) "Evolution du coefficient de dilatation thermique du beton en fonction de sa maturite aux tout premiers ages", *Materials and Structures*, Vol. 27, pp. 596–605.

Larive, C. & Gremillon, K. (2007) "Certification of shotcrete nozzlemen around the world", *Underground Space – the 4th Dimension of Metropolises*, Bartak, Hrdina, Romancov & Zlamal (eds.), Taylor & Francis, London, pp. 1395–1400.

Larive, C., Rogat, D., Chamoley, D., Regnard, A. & Pannetier, T. (2016) "Influence of fibres on the creep behaviour of reinforced sprayed concrete", *Proceedings of the World Tunnel Congress 2016*, San Francisco, CA.

Leca, E. & Dormieux, L. (1990) "Face stability of tunnels in frictional soils", *Geotechnique*, Vol. 40, No. 4, pp. 581–606.

Lee, K.M. & Rowe, R.K. (1989) "Deformations caused by surface loading and tunnelling: The role of elastic anisotropy", *Geotechnique*, Vol. 39, No. 1, pp. 125–140.

Lehto, J. & Harbron, R. (2011) "Achieving the highest standards – through the EFNARC Nozzleman Certification Scheme", *6th International Symposium on Sprayed Concrete*, Tromso, pp. 246–256.

Lootens, D., Oblak, L., Lindlar, B. & Hansson, M. (2014) "Ultrasonic Wave Propagation for Strength Measurements: Application in Shotcrete", *7th International Symposium on Sprayed Concrete*, Sandefjord, pp 287–293.

Lukas, W., Huber, H., Kusterle, W., Pichler, W., Testor, M. & Saxer, A. (1998) "Bewertung von neuentwickelten Spritzbetonverfahrenstechniken", *Strassenforschung*, Heft 474, Vienna.

MacDonald, M. (1990) "Heathrow Express Rail Link – Geotechnical Design Parameter Report", MMC/01/23/R/1508, December 1990.

MacKay, J. & Trottier, J.-F. (2004) "Post-crack creep behaviour of steel and synthetic FRC under flexural loading", *Shotcrete: More Engineering Developments*, Bernard (ed.), A. A. Balkema, Leiden, pp. 183–192.

Mair, R.J. (1993) "Unwin Memorial Lecture 1992: Developments in geotechnical engineering research: Application to tunnels and deep excavations", *Proceedings Institution of Civil Engineers, Civil Engineering*, 1993, Vol. 93, February, pp. 27–41.

Mair, R.J. (1998) "Geotechnical aspects of design criteria for bored tunnelling in soft ground", *Tunnels & Metropolises*, Negro Jr & Ferreira (eds.), Taylor & Francis, London, pp. 183–199.

Melbye, T.A. (2005) *Sprayed Concrete for Rock Support*, UGC International, Degussa, Zurich.

Meschke, G. (1996) "Elasto-viskoplastische Stoffmodelle für numerische Simulationen mittels der Methode der Finiten Elemente", Habilitationsschrift, TU Wien.

Michelis, P. (1987) "True triaxial cyclic behaviour of concrete and rock in compression", *International Journal of Plasticity*, Vol. 3, pp. 249–270.

Mills, P., Tadolini, S. & Thomas, A.H. (2018) "Ultra fast rapid hardening sprayed concrete", *8th International Symposium on Sprayed Concrete*, Trondheim, pp. 238–248.

Minh, N.A. (1999) "The investigation of geotechnical behaviour near excavated tunnel face by means of three-dimensional stress-flow coupled analysis", MEng Dissertation, Asian Institute of Technology, Bangkok.

Morgan, D.R., Zhang, L. & Pildysh, M. (2017) "New Hemp-based Fiber Enhances Wet Mix Shotcrete Performance", *Shotcrete*, Spring 2017, pp. 36–45.

Moussa, A.M. (1993) "Finite element modelling of shotcrete in tunnelling", PhD thesis, University of Innsbruck.

Mosser, A. (1993) "Numerische Implementierung eines zeitabhangigen Materialgesetzes für jungen Spritzbeton in Abaqus", Diplomarbeit, Montanuniversitat Leoben.

Muir Wood, A.M. (1975) "The circular tunnel in elastic ground", *Geotechnique*, Vol. 25, No. 1, pp. 115–127.

Myrdal, R. (2011) "Chemical reflections on accelerators for sprayed concrete: Past, present and future challenges", *6th International Symposium on Sprayed Concrete*, Tromso, pp. 304–316.

Myrdal, R. & Tong, S. (2018) "Sprayed concrete without Portland Cement", *8th International Symposium on Sprayed Concrete*, Trondheim, pp. 249–256.

Myren, S.A. & Bjontegaard, O. (2014) "Fibre reinforced sprayed concrete (FRSC): Mechanical properties and pore structure characteristics", *7th International Symposium on Sprayed Concrete*, Sandefjord, pp. 305–313.

NCA (1993) Sprayed concrete for rock support, Publication No. 7, Norwegian Concrete Association.

NCA (2011) Sprayed concrete for rock support, Publication No. 7, Norwegian Concrete Association.

Negro, A., Kochen, R., Goncalves, G.G., Martins, R.M. & Pinto, G.M.P. (1998) "Prediction and measurement of stresses in sprayed concrete lining (Brasilia South Wing tunnels)", *Tunnels and Metropolises*, Negro Jr & Ferreira (eds.), Taylor & Francis, London, pp. 405–410.

Neville, A.M. (1995) *"Properties of Concrete"*, Addison Wesley Longman Ltd, Harlow.

Neville, A.M., Dilger, W.H. & Brooks, J.J. (1983) *Creep of Plain & Structural Concrete*, Construction Press (Longman), Harlow.

NFF (2011) "Rock mass grouting in Norwegian tunnelling", Norwegian Tunnelling Society (NFF) Publication No. 20, 2011.

Niederegger, C. & Thomaseth, D. (2006) "Strength properties of shotcrete: Influence of stickiness of the wet mix", *Tunnel*, Vol. 1, 2006, pp. 14–25, 2006.

Nordström, E. (2001) "Durability of steel fibre reinforced shotcrete with regard to corrosion", *Shotcrete: Engineering Developments*, Bernard (ed.), Swets & Zeitlinger, Lisse, pp. 213–217.

Nordström, E. (2016) "Evaluation after 17 years with field exposures of cracked steel fibre reinforced shotcrete", BeFo Report 153, Rock Engineering Research Foundation, ISSN: 1104-1773.

Norris, P. (1999) "Setting the ground rules for wet-mix sprayed concrete", *Concrete*, May 1999, pp. 16–20.

Norris, P. & Powell, D. (1999) "Towards quantification of the engineering properties of steel fibre reinforced sprayed concrete", *3rd International Symposium on Sprayed Concrete*, Gol, Norway, pp. 393–402.

Oberdörfer, W. (1996) "Auswirkung von unterschiedlichen Betonnachbehandlungsmassnahmen auf die Qualitat des Nassspritzbetons", Diplomarbeit, University of Innsbruck.

ÖBV (1998) "Guideline on shotcrete", Österreichischer Beton Verein.

ÖBV (2013) "Guideline on shotcrete", Österreichischer Beton Verein. http://www.bautechnik.pro.

Oliveira, D.A.F. & Paramaguru, L. (2016) "Laminated rock beam design for tunnel support", *Australian Geomechanics*, Vol. 51, No. 3, September 2016.

Owen, D.J.R. & Hinton, E. (1980) *"Finite Elements in Plasticity: Theory and Practice"*, Pineridge Press Ltd, Swansea.

Palermo, G. & Helene, P.R.d.L. (1998) "Shotcrete as a final lining for tunnels", *Tunnels and Metropolises*, Negro Jr & Ferreira (eds.), Taylor & Francis, London, pp. 349–354.

Panet, M. & Guenot, A. (1982) "Analysis of convergence behind the face of a tunnel", *Tunnelling '82*, IMM, pp. 197–204.

Papanikolaou, I., Davies, A., Jin, F., Litina, C. & Al-Tabbaa, A. (2018) "Graphene oxide/cement composites for sprayed concrete tunnel linings", *World Tunnel Congress 2018*, Dubai.

Penny, C., Stewart, J., Jobling, P.W. & John, M. (1991) "Castle Hill NATM tunnels: Design and construction", *Tunnelling '91*, IMM, pp. 285–297.

Pichler, P. (1994) "Untersuchungen zum Materialverhalten und Überprüfungen vom Rechenmodellen für die Simulation des Spritzbetons in Finite-Elemente-Berechnungen", Diplomarbeit, Montanuniversitat Leoben.

Plizzari, G. & Serna, P. (2018) "Structural effects of FRC creep", *Materials and Structures*, 2018, pp. 51–167.

Podjadtke, R. (1998) "Bearing capacity and sprayed-in-behaviour of the star profile compared to other lining profiles", *Tunnel*, Vol. 2, 1998, pp. 46–51.

Pöttler, R. (1985) "Evaluating the stresses acting on the shotcrete in rock cavity constructions with the Hypothetical Modulus of Elasticity", *Felsbau*, Vol. 3, No. 3, pp. 136–139.

Pöttler, R. (1990) "Green shotcrete in tunnelling: Stiffness – strength – deformation", *Shotcrete for Underground Support VI*, pp. 83–91.

Pöttler, R. (1993) "To the limits of shotcrete linings", *Spritzbeton Technologie 3rd International Conference*, pp. 117–128.

Pöttler, R. & Rock, T.A. (1991) "Time Dependent Behaviour of Shotcrete and Chalk Mark, Development of a Numerical Model", *ASCE* Geotechnical Congress Engineering in Boulder, Colorado, ASCE, Reston, pp. 1319–1330.

Pound, C. (2002) – personal communication.

Powell, D.B., Sigl, O. & Beveridge, J.P. (1997) "Heathrow Express – design and performance of platform tunnels at Terminal 4", *Tunnelling '97*, IMM, pp. 565–593.

Powers, T.C. (1959) "Causes and control of volume change", *Journal of the PCA Research & Development Laboratories*, Vol. 1, January 1959, pp. 30–39.

Probst, B. (1999) "Entwicklung einer Langzeitdruckversuchsanlage für den Baustellenbetrieb zur Bestimmung des Materialverhaltens von jungem Spritzbeton", Diplomarbeit, Montanuniversitat Leoben.

Purrer, W. (1990) "Spritzbeton in den NOT-Abschnitten des Kanaltunnels", *Spritzbeton Technologie '90*, Universität pp. 67–78.

Rabcewicz, L.V. (1969) "Stability of tunnels under rock load Parts 1–3", *Water Power*, Vol. 21 Nos. 6–8, pp. 225–229, 266–273, 297–304.

Rathmair, F. (1997) "Numerische Simulation des Langzeitverhaltens von Spritzbeton und Salzgestein mit der im FE Program Abaqus implementierten Routine", Diplomarbeit, Montanuniversitat Leoben.

RILEM (2003) "TC 162-TDF, Test and design methods for SFRC", *Materials and Structures*, Vol. 36, 2003.

Rispin, M., Kleven, O.B., Dimmock, R. & Myrdal, R. (2017) "Shotcrete: Early strength and re-entry revisited – practices and technology", *Proceedings of the First International Conference on Underground Mining Technology*, Hudyma & Potvin (eds.), Australian Centre for Geomechanics, Perth, pp. 55–70.

Rokahr, R.B. & Lux, K.H. (1987) "Einfluss des rheologischen Verhaltens des Spritzbetons auf den Ausbauwiderstand", *Felsbau*, Vol. 5, No. 1, pp. 11–18.

Rokahr, R.B. & Zachow, R. (1997) "Ein neues Verfahren zur täglichen Kontrolle der Auslastung einer Spritzbetonschale", *Felsbau*, Vol. 15, No. 6, pp. 430–434.

Rose, D. (1999) "Steel-fiber-reinforced-shotcrete for tunnels: An international update", *Rapid Excavation and Tunnelling Conference Proceedings 1999*, Orlando, FL, pp. 525–536.

Röthlisberger, B. (1996) "Practical experience with the single-shell shotcrete method using wet-mix shotcrete during the construction of the Vereina Tunnel", *Spritzbeton Technologie '96*, pp. 49–56.

Rowe, R.K. & Lee, M.K. (1992) "An evaluation of simplified techniques for estimating three-dimensional undrained ground movements due to tunnelling in soft soils", *Canadian Geotechnical Journal*, Vol. 29, pp. 31–59.

Rudberg, E. & Beck, T. (2014) "Pozzolanic material activator", *7th International Symposium on Sprayed Concrete*, Sandefjord, pp. 336–341.

Ruzicka, J., Kochanek, M. & Vales, V. (2007) "Experiences from driving dual-rail tunnels on Prague metro's IV C2 route", *Underground Space – the 4th Dimension of Metropolises*, Bartak, Hrdina, Romancov & Zlamal (eds.), Taylor & Francis, London, pp. 1055–1062.

Sandbakk, S., Miller, L.W. & Standal, P.C. (2018) "MiniBars – A new durable composite mineral macro fiber for shotcrete, meeting the energy absorption criteria for the industry", *8th International Symposium on Sprayed Concrete*, Trondheim, pp. 272–282.

Schiesser, K. (1997) "Untersuchung des Langzeitverhaltens permanenter Spritzbetonauskleidung im Tunnelbau mittels numerischer Simulation", Diplomarbeit, Montanuniversitat Leoben.

Schorn, H. (2004) "Strengthening of reinforced concrete structures using shotcrete", *Shotcrete: More Engineering Developments*, Bernard (ed.), A. A. Balkema, Leiden, pp. 225–232.

Schmidt, A., Bracher, G. & Bachli, R. (1987) "Erfahrungen mit Nassspritzbeton", *Schweizer Baublatt*, Vol. 59, No. 60, pp. 54–60.

Schröpfer, T. (1995) "Numerischer Analyse des Tragverhaltens von Gebirgsstrecken mit Spritzbetonausbau im Ruhrkarbon", PhD Thesis, TU Clausthal.

Schubert, P. (1988) "Beitrag zum rheologischen Verhalten von Spritzbeton", *Felsbau*, Vol. 6, No. 3, pp. 150–153.

Seith, O. (1995) "Spritzbeton bei hohen Temperaturen", *Internal Report*, ETH Zurich.

Sercombe, J., Hellmich, C., Ulm, F.-J. & Mang, H. (2000) "Modeling of early-age creep of shotcrete. I: Model and model parameters", *Journal of Engineering Mechanics*, Vol. 126, No. 3, pp. 284–291.

Sharma, J.S., Zhao, J. & Hefny, A.M. (2000) "NATM – Effect of shotcrete setting time and excavation sequence on surface settlements", *Tunnels and Underground Structures*, Zhao, Shirlaw & Krishnan (eds.), Taylor & Francis, London, pp. 535–540.

Matya, E.S., Suarez Diaz, J., Vivier, R., Marchand, E. & Ahmad, S. (2018) "Design and Construction of Inclined Escalator Shafts and Stair Adit at Liverpool St and Whitechapel Stations", Crossrail Learning Legacy, https://learninglegacy.crossrail.co.uk/.

Sjolander, A., Hellgren, R. & Ansell, A. (2018) "Modelling aspects to predict failure of a bolt-anchored fibre reinforced shotcrete lining", *8th International Symposium on Sprayed Concrete*, Trondheim, pp. 283–297.

Smith, K. (2016) "Where next for SCL?", *Tunnelling Journal*, September 2016, pp. 28–33.

Smith, K. (2018) "Stemming the flow", *Tunnelling Journal*, February/March 2018, pp. 30–34.

Soliman, E., Duddeck, H. & Ahrens, H. (1994) "Effects of development of stiffness on stresses and displacements of single and double tunnels", *Tunnelling and Ground Conditions*, Abdel Salam (ed.), A. A. Balkema, Rotterdam, pp. 549–556.

Spirig, C. (2004) "Sprayed concrete systems in the Gotthard base tunnel", *Shotcrete: More Engineering Developments*, Bernard (ed.), A. A. Balkema, Leiden, pp. 245–249.

Springenschmid, R., Schmiedmayer, R. & Schöggler, G. (1998) "Comparative examination of shotcrete with gravel or chippings as aggregate", *Tunnel*, Vol. 2, 1998, pp. 38–45.

Stärk, A. (2002) "Standsicherheitsanalyse von Sohlsicherungen aus Spritzbeton", *Forschungsergebnisse aus dem Tunnel- und Kavernenbau*, Heft 22, Universität Hannover, Institut für Unterirdisches Bauen, Hannover.

Stärk, A., Rokahr, R.B. & Zachow, R. (2002) "What do we measure: Displacements or safety?", *Proceedings of the 5th North American Rock Mechanics Symposium and the 17th Tunnelling Association of Canada Conference: NARMS-TAC 2002 "Mining and Tunnelling Innovation and Opportunity"*, Toronto, University of Toronto Press.

Steindorfer, A.F. (1997) "Short-term prediction of rock mass behaviour in tunnelling by advanced analysis of displacement monitoring data", PhD Thesis, TU Graz.

Stelzer, G. & Golser, J. (2002) "Untersuchungen zu geometrischen Imperfektionen von Spritzbetonschalen – Ergebnisse aus Modellversuchen und numerischen Berechnungen", *Spritzbeton Technologie 2002*, Alpbach, pp. 105–111.

Strobl, B. (1991) "Die NATM im Boden in Kombination mit Druckluft", Diplomarbeit, Universität für Boden Kultur (BOKU), Vienna.

Strubreiter, A. (1998) "Wirtschaftlichkeitsvergleich von verschiedenen Spritzbetonverfahren im Tunnelbau", Diplomarbeit, University of Innsbruck.

Su, J. & Bloodworth, A.G. (2014) "Experimental and numerical investigation of composite action in composite shell linings", *7th International Symposium on Sprayed Concrete*, Sandefjord, pp. 375–386.

Su, J. & Bloodworth, A.G. (2016) "Utilizing composite action to achieve lining thickness efficiency for Sprayed Concrete Lined (SCL) tunnels", *Proceedings of the World Tunnel Congress 2016*, San Francisco, CA.

Su, J. & Bloodworth, A. (2018) "Numerical calibration of mechanical behaviour of composite shell tunnel linings", *Tunnelling and Underground Space Technology*, Vol. 76, pp. 107–120.

Su, J. & Uhrin, M. (2016) "Interactions in sprayed waterproof membranes". *Tunnelling Journal*, October/November 2016, pp. 20–27.

Swoboda, G. & Moussa, A.M. (1992) "Numerical modeling of shotcrete in tunnelling", *Numerical Models in Geomechanics*, Pande & Pietruszczak (eds.), Taylor & Francis, London, pp. 717–727.

Swoboda, G. & Moussa, A.M. (1994) "Numerical modeling of shotcrete and concrete tunnel linings", *Tunnelling and Ground Conditions*, Abdel Salam (ed.), A. A. Balkema, Rotterdam, pp. 427–436.

Swoboda, G., Moussa, A.M., Lukas, W. & Kusterle, W. (1993) "On constitutive modelling of shotcrete", *International Symposium on Sprayed Concrete*, Kompen, Opsahl & Berg (eds.), Norwegian Concrete Institute, Fagernes, Norway, pp. 133–141.

Swoboda, G. & Wagner, H. (1993) "Design based on numerical modelling. A requirement for an economical tunnel construction", *Rapid Excavation and Tunnelling Conference Proceedings 1993*, pp. 367–379.

Szechy, K. (1973) *The Art of Tunnelling*, Akademiai Kiado, Budapest.

Tatnall, P.C. & Brooks, J. (2001) "Developments and applications of high performance polymer fibres in shotcrete", *Shotcrete: Engineering Developments*, Bernard (ed), Swets & Zeitlinger, Lisse, pp. 231–235.

Testor, M. (1997) "Alkaliarme Spritzbetontechnologie - Verfahrenstechnik; Druck-festigkeits, Ruckprall- und Staubuntersuchungen", PhD Thesis, University of Innsbruck.

Testor, M. & Kusterle, W. (1998) "Ermittlung der Spritzbetondruckfestigkeiten – Modifiziertes Setzbolzenverfahren und Abhängigkeit der Druckfestigkeit von der Probekörpergeometrie", *Zement und Beton 3* Vol. 98, Vereinigung der Österreichischen Zementindustrie, Vienna, pp. 20–23.

Testor, M. & Pfeuffer, M. (1999) "Staub- und Rückprallreduktion beim Auftrag von Trockenspritzbeton", *Spritzbeton-Technologie '99*, Innsbruck-Igls, pp. 137–149.

Thomas, A.H. (2003) "Numerical modelling of sprayed concrete lined (SCL) tun-nels", PhD Thesis, University of Southampton.

Thomas, A.H. (2014) "Design methods for fibre reinforced concrete", *Tunnelling Journal*, June/July, pp. 44–48.

Thomas, A.H., Casson, E.M. & Powell, D.P. (2003) "Common ground – the inte-gration of the design and construction of a sprayed concrete lined (SCL) tun-nel in San Diego, USA", *Underground Construction 2003*, IMMM/BTS, pp. 71–82.

Thomas, A.H. & Dimmock, R. (2018) "Design philosophy for permanent sprayed concrete linings", *8th International Symposium on Sprayed Concrete*, Trondheim, pp. 298–312.

Thomas, A.H. & Pickett, A.P. (2012) "Where are we now with sprayed concrete lining in tunnels?", *Tunnelling Journal*, April/May, pp. 30–39.

Thomas, A.H., Powell, D.B. & Savill, M. (1998) "Controlling deformations during the construction of NATM tunnels in urban areas", *Underground Construction in Modern Infrastructure*, Franzen, Bergdahl & Nordmark (eds.), A. A. Balkema, Rotterdam, pp. 207–212.

Thring, L., Jung, H.-I. & Green, C. (2018) "Developing a simple spring interface modelling technique for a composite lining bond interface", *World Tunnel Congress 2018*, Dubai.

Thumann, M. & Kusterle, W. (2018) "Pumpability of wet mix sprayed concrete with reduced clinker content", *8th International Symposium on Sprayed Concrete*, Trondheim, pp. 313–324.

Thumann, M., Saxer, A. & Kusterle, W. (2014) "Precipitations in the tunnel drain-age system – the influence of sprayed concrete and other cement bound mate-rials", *7th International Symposium on Sprayed Concrete*, Sandefjord, pp. 375–386.

Triclot, J., Rettighieri, M. & Barla, G. (2007) "Large deformations in squeez-ing ground in the Saint-Martin La Porte gallery along the Lyon-Turin Base Tunnel", *Underground Space – the 4th Dimension of Metropolises*, Bartak, Hrdina, Romancov & Zlamal (eds.), Taylor & Francis, London, pp. 1093–1097.

Trottier, J.-F., Forgeron, D. & Mahoney, M. (2002) "Influence of construction joints on the flexural performance of fibre and welded wire mesh reinforced wet mix shotcrete panels", *4th International Symposium on Sprayed Concrete*, Davos, pp. 11–25.

Tyler, D. & Clements, M.J.K. (2004) "High toughness shotcrete for large deformation control at Perseverance Mine", *Shotcrete: More Engineering
Developments*, Bernard (ed.), A. A. Balkema, Leiden, pp. 259–266.

Ullah, S., Pichler, B., Scheiner, S. & Hellmich, C. (2011) "Composition-related sensitivity analysis of displacement loaded sprayed concrete tunnel shell, employ-
ing micromechanics and thin shell theory", *6th International Symposium on
Sprayed Concrete*, Tromso, pp. 203–211.

Ulm, J. & Coussy, O. (1995) "Modeling of thermochemomechanical couplings of
concrete at early ages", *Journal of Engineering Mechanics*, Vol. 121, No. 7,
July 1995, pp. 785–794.

Ulm, J. & Coussy, O. (1996) "Strength growth as chemo-plastic hardening in
early-age concrete", *Journal of Engineering Mechanics*, Vol. 122, No. 12,
December 1996, pp. 1123–1133.

Van der Berg, J.P. (1999) "Measurement and prediction of ground movements
around three NATM tunnels", PhD Thesis, University of Surrey.

Vandewalle, M. (1996) *Dramix: Tunnelling the World*, Zwevegem, Belgium, N. V.
Bekaert SA.

Vandewalle, M., Rock, T., Earnshaw, G. & Eddie, C. (1998) "Concrete reinforcement with steel fibres", *Tunnels & Tunnelling International*, April, pp. 39–41.

Varley, N. & Both, C. (1999) "Fire protection of concrete linings in tunnels",
Concrete, May, pp. 27–30.

Vogel, F., Sovjak, R. & Peskova, S. (2017) "Static response of double shell concrete lining with a spray-applied waterproofing membrane", *Tunnelling and
Underground Space Technology*, Vol. 68, pp. 106–112.

Ward, W.H., Tedd, P. & Berry, N.S.M. (1983) "The Kielder experimental tunnel:
Final results", *Geotechnique*, Vol. 33, No. 3, pp 275–291.

Watson, P.C., Warren, C.D., Eddie, C. & Jager, J. (1999) "CTRL North Downs
Tunnel", *Tunnel Construction & Piling '99*, IMM, pp. 301–323.

Weber, J.W. (1979) "Empirische Formeln zur Beschreibung der Festigkeitsentwick-
lung und der Entwicklung des E-Moduls von Beton", *Betonwerk- & Fertig-
teiltechnik*, Vol. 12, pp. 753–756.

Wetlesen, T. & Krutrok, B. (2014) "Measurement of shotcrete thickness in tunnel with Bever 3D laser scanner operated from the robot", *7th International
Symposium on Sprayed Concrete*, Sandefjord, pp. 401–417.

Winterberg, R. & Dietze, R. (2004) "Efficient passive fire protection systems for
high performance shotcrete", *Shotcrete: More Engineering Developments*,
Bernard (ed.), A. A. Balkema, Leiden, pp. 275–290.

Wittke, W. (2007) "New high-speed railway lines Stuttgart 21 and Wendlingen-
Ulm – approximately 100 km of tunnels", *Underground Space – the 4th
Dimension of Metropolises*, Bartak, Hrdina, Romancov & Zlamal (eds.),
Taylor & Francis, London, pp. 771–778.

Wittke-Gattermann, P. (1998) *"Verfahren zur Berechnung von Tunnels in
quellfähigem Gebirge und Kalibrierung an einem Versuchsbauwerk"*, Verlag
Glückauf, Essen, 1998.

Wong, R.C.K. & Kaiser, P.F. (1988) "Design and performance evaluation of vertical shafts: Rational shaft design method and verification of design method",
Canadian Geotechnical Journal, Vol. 25, pp. 320–337.

Wong, R.C.K. & Kaiser, P.F. (1989) "Design and performance of vertical shafts", *Canadian Tunnelling 1989*, pp. 73–87.

Woods, R.I. & Clayton, C.R.I. (1993) "The application of the CRISP finite element program to practical retaining wall problems", *Proceedings of the ICE conference on Retaining Structures*, Thomas Telford, London, pp. 102–111.

Yin, J. (1996) "Untersuchungen zum zeitabhangigen Tragverhalten von tiefliegenden Hohlraumen im Fels mit Spritzbetonausbau", PhD thesis, TU Clausthal.

Yun, K.-K., Eum, Y.-D. & Kim, Y.-G. (2011) "Effect of crushed sand gradation on the rheology properties of high performance wet-mix shotcrete", *6th International Symposium on Sprayed Concrete*, Tromso, pp. 441–452.

Yun, K.-K., Eum, Y.-D. & Kim, Y.-G. (2014) "Chloride penetration resistance of shotcrete according to mineral admixture type and supplemental ratio", *7th International Symposium on Sprayed Concrete*, Sandefjord, pp. 418–428.

Yun, K.-K., Han, S.-Y., Lee, K.-R. & Kim, Y.-G. (2018) "Applications of cellular sprayed concrete at tunnel portals", *8th International Symposium on Sprayed Concrete*, Trondheim, pp. 325–334.

Yurdakul, E. & Rieder, K.-A. (2018) "The importance of rheology on shotcrete performance", *8th International Symposium on Sprayed Concrete*, Trondheim, pp. 335–345.

Zangerle, D. (1998) "The use of wet mix sprayed concrete", *Tunnels and Metropolises*, Negro Jr & Ferreira (eds.), Taylor & Francis, London, pp. 861–867.

Index

<cite/>

Printed in the United States
by Baker & Taylor Publisher Services